POPULATION AND COMMUNITY ECOLOGY

Encyclopaedia of Environmental Pollution and Awareness in 21st Century Series

POPULATION AND COMMUNITY ECOLOGY

Edited by

V.K. Prabhakar

ANMOL PUBLICATIONS PVT. LTD.

NEW DELHI - 110 002 (INDIA)

ANMOL PUBLICATIONS PVT. LTD.
4374/4B, Ansari Road, Daryaganj
New Delhi - 110 002

Population and Community Ecology

First Edition, 2001

ISBN 81-261-0923-8

PRINTED IN INDIA

Published by J.L. Kumar for Anmol Publications Pvt. Ltd., New Delhi - 110 002 and Printed at Mehra Offset Press, Delhi.

Contents

PREFACE

Today, the cry of 'pollution' is heard from all the nooks and corners of the world, and pollution has become a major threat to the very existence of mankind on this earth. It is the major challenge of our times.

The pollution of various resources has gone to such an extent that we are unable to breathe fresh air and drink fresh water. On one hand the advancements of science and technology have added to the human comforts by giving us automobiles, electrical appliances, supersonic jets, space crafts, better medicines, better chemicals to control harmful insects and other pests, etc., but, on the other hand, they have given us a very serious problem to face that is pollution. Pollution can be defined as an undesirable change in the physical, chemical, or biological characteristics of the air, water, or land that can harmfully affect health, survival, or activities of humans or other living organisms.

Pollution does not have to cause physical harm, but it may merely interfere with human activities. Chemicals spewed into the air or water from an industrial plant may be harmful to humans and other organisms living near the plant. If expensive pollution controls are required, the plant may be forced to shut down.

India faces serious problems of environmental pollution. Water pollution from untreated community and industrial wastes flowing into our rivers and streams, increasing use of fertilizers and pesticides, dumping of organic wastes etc., are issues that need to be tackled on the priority. Air pollution, long regarded as the bane of industrialised nations looms large over our country.

Here, we are presenting this *Encyclopaedia of Environmental Pollution and Awareness in Twenty-first Century* in fifty volumes with a view to make people conscious about the seriousness of pollution in the present time and its remedies.

Editor

1 Ecology of Populations

Population is an interacting, or potentially interacting group of individuals usually of the same species. It may be the population of ants in an ant hill, the population of salmon in a river, the population of a deer in a state and so forth. Biologically, a population should have some natural boundaries as is the ants in an ant hill, but the term is often used with artificial boundaries in mind, as in referring to the population of deer in a state. A state might have several biological population of deer, but for economic or political consideration it is convenient to put them together. The human population representing varied ethnic religious or economic groups these population within one nation might have different characteristics.

In the initial study of populaion it is helpful to recognize certain attribute of population which distinguish them from individual organisms like organisms have a structural organization, a functional unity, and a pattern of growth and development unlike individuals they have group attributes and statistical properties that no single individual possess.

The proportion of individuals of different ages in the group. Genetic characteristics especially applicable to population ecology as for example, adaptiveness reproductive fitness and persistence. Individual organisms are born, have age, grow and die but such characteristic as birth rate, death rate, density and other listed above are meaningful only at the group level. If we are to understand thoroughly the ecology of a species, we must study and measure these population group characteristics as well as know the life history and identifying features of the species. And of course, we all must be concerned about the vital statistics of human population.

Some of the most important aspects of population structure are population size, density, spatial distribution age and sex rates, breeding structure and social organization. Population sizes may vary from a few individuals to millions of individuals. The population of whooping cranes (Grus American) numbers about 50 individuals a single group of birds barely clining to existence. Other rare and endangered species around the world also represent very small populations.

The world's population of the Indian lion (Panthera Ieo) numbers less than 200 individuals living in the Gir forest of western India. The California condor (Gymonogyps californicus), golden marmoset (Leontideus rosalia) Javan rhinoceros (Didermocerus sumatrensis) orangutan (Pongo pygmaeus), Gray jungle fowl (Gallus sonnerati) and many other species are now represented by small remnant populations wavering on the edge between life and death. At the other extreme, the world contains many species of tremendous abundance.

A population of starlings in a sigle winter flocking roost in eastern United States may number 5 million birds. The white-tailed deer population of Pennsylvania was estimated in 1946 to contain 1 million deer; the rhesus monkey population of one province in northern India was estimated in 1960 to contain between 800,000 and 1 million monkeys (it now probably contains less than half that number); the Norway rat population of Baltimore was estimated in World War II at 400,000; and the bison population of the Great Plains was estimated to contain 60 million individuals in the early nineteenth century, though by the 1890's the bison population had crashed to less than 600 through the greed and disruptive influence of man.

Population estimates

These figures raise the question; of course as to how such data are obtained, and this is a large and complicated subject in itself. Censusing wild animal population is fraught with many difficulties of sampling error and statistical bias. Occasionally direct counts are possible as in aerial photographs of big game herds or sea bird colonies, but for most animals this is impossible and various kinds of sampling methods must be employed. Most animals are not readily visible because of their behavior and habitat, or because they exist in such abundance or scarcity that they cannot be readily counted. It therefore becomes necessary to estimate numbers through programs of capture and resembling. With many animals systematic program of capture-mark-recapture has been helpful in obtaining estimates of abundance.

The ratio of marked to unmarked animals in subsequent trapping runs provides a population estimate known as the Lincoln Index. If, for example one is studying a resident bird or rodent population in a forest, and succeeds in capturing and banading 100 individuals all of which are released, and then if on a second trapping program, one again catches 100 animals of which 20 were previously banded one would estimate the total original population at 500 animals.

The validity of the method involves several major assumptions: (1) that the marked animals mix randomly in the population, and (2) that the probability of recapture is the same for each individual regardless of whether or not it is marked, and (3) that there be no immigration or emigration;, death or births between the sampling periods. Actually, these conditions are rarely attained in natural populations. Many studies on birds and mammals have shown that some individuals are easily captured whereas others are rarely captured. There are thus "trap-happy" individuals and "trap shy" individuals.

In some species, one experience with trapping on the part of an individual animal may reduce the probability of subsequent recapture, whereas in others it may enhance it. Such behavior patterns bias the data and reduce the value of the population estimates made with this method. Also, birth, death, immigration and emigration are often continuous processes within populations, so their effect can rarely be eliminated entirely. Nonetheless, the Lincoln Index approaches validity in certain cases. It can be a helpful tool for population estimates, particularly in the following cases: (1) if a capture method is available which is relatively nonselective and random in its action, (2) if a relatively confined population is available so that emigration and immigration do not occur, as an island population, a woodland population surrounded by forest, an oasis , etc., (3) if the population can be sampled at a time when it is relatively stable, not reproducing during the sampling period, and not migrating. Under these circumstances, capture-mark-recapture technique can produce the best available result on certain animal populations.

Another method of estimating animal population size is based on a change in sex ration before and after a known harvest of one sex has been removed. This has been called the Kelker Ratio, and has been used in the estimation of deer, pheasants, and other wildlife populations where hunting statistics are available. In a study of California deer by Dasmann (1952), the sex ratio observed before the hunting season was 74 bucks to 139 does, and after the hunting season in the same area it was 50 bucks to 139 does. The harvest in the hunting season was known to be 246 bucks. Hence the removal of 246 bucks population should be in the same proportion to the original sex ratio as the total buck harvest was to the change in sex ratios produced by the harvest.

One can readily see some of the observational problems and potential biases in the Kelker Ratio as the Lincoln Index. The Kelker Ratio assumes that the observations of sex ratios before and after harvests are equivalent. It the buck become shyer than the does and are harder to observe after the hunt, this enters a major bias. It is essential that the field ecologist be fully experienced and knowledgeable in the habits of the animal he is studying before he applies these population estimation techniques. He can then recognize the sampling problems and possible sources of error. If reasonably unbiased sample estimates are obtained, it is then important to subject them to statistical treatments such as standard error calculations, or analyses of variance to evaluate the sampling errors which might occur by chance alone. It is thus important for ecologist and statisticians to work closely together in population studies.

Population density is an important aspect of both population size and distribution. In many populations, the precise boundaries of the population are unknown. Though the distributional limits of the species may be recognized each species usually consists of several populations. The essential feature which distinguishes a population is whether or not the individuals are potentially interactive, or potentially interbreeding. If a species consists of two or more discrete aggregations, which do not interact or influence each other these aggregations can then be considered populations. If this condition persists over an extended period of time, different races, subspecies and even species may emerge as gene frequencies change and evolution occurs.

Since populational boundaries are often unknown the most practical description of population abundance is frequently in terms of densities. As an example many ecologic studies of small rodents have shown population densities usually in the range of 0.5 to 20 animals per acre. But (1940) showed population densities of the deer mouse (Peromyscus leucopus)in southern Michigan woodlots to range from 3.08 to 10.32 individuals per acre, Blair (1951) estimated the population density of beach mice or old field mice (Peromyscus polionotus) to range from 0.83 to 1.41 individuals per acre. Ashby estimated population densities of the bank vote (Clethrionomy glaeolus) a relative of the meadow mouse, to run as high as 17 to 18 individuals per acre in central England. An important consideration, which will be emphasized later, is that many populations fluctuate widely in density and size in the course of natural events. Occasionally small mammal populations reach amazingly high densities.

In California, Krebs described a meadow mouse or vole population (Microtus Californicus) which achieved a density of 150 to 300 individuals per acre in a relatively small area, and Parson described a population of house mice in California which increased from a density of less than 5 mice per acre to 300 mice per acre within 6 months. At these high levels, or course, rodents can become agricultural problem and can precipitate economic losses of crops or stored food products. Splitte has estimated that wild rats (Bandicota bengalensis) in

Calcutta grain storage warehouses may achieve densities of one rat per sequare meter, equivalent to 4,000 per acre. These represent extreme concentrations, or course, but they indicate the population potential of some species.

Another important consideration in concepts of both population size and density is the spatial distribution of individuals. Spatial distribution involves a fascinating interplay of behavioral and ecologic factors, and is a prime example of the inseparable tie between behavior and ecology. We have been speaking of population densities as though animals were randomly distributed. This is rarely true, and it is much more common for population to be clumped and nonrandomly distributed.

A typical pattern of spatial distribution within a population might accurately portray the distribution of insects in an old field, isopods or millipedes in aforest, nesting birds or prairie dogs in a grassland, baboons in a savannah, fish in a lake, or even plankton in the sea. The reason for such clumping or nonrandom distribution might be numerous. In some cases clumping would be attributable to the heterogeneity of the environment the distribution of food, cover and shelter. For example, dung beetles in an old field would be clumped around the dung isopods and millipedes around the fallen log, etc. In other cases, the clumping might be due to the social behavior. This clumping may be related to environmental resources, but not necessarily so.

This is not to say that random dispersion of animals never occurs. Park pointed out that flour beetles (Tribalism confusum) are distributed in flour according to a Poison distribution, indicating that the individuals are spaced at random. Cole found that certain spiders were distributed randomly under boards placed on the forest floor. These examples are probably exceptions to the general rule, and they are most likely attributable to either the homogeneous or artificial nature of the environment in each case. For purposes of generalization it can be stated with reasonable assurance that most animals and many plants in natural environments show a clumped or nonrandom pattern of distribution.

Sex Ratios and Age Structures

It is obvious that animals within a population differ in a number of ways: sex, age, breeding condition health, physical condition and social status. Demography, or the numerical analysis of populations, is concerned with the sex, age and breeding condition of the population.

Medical science and epistemology are primarily concerned with the health and physical condition of the population and behavioural science is primarily concerned with the social and behavioral states within the population. In most vertebrate populations primary sex ratios at hatching or birth approach 50 percent male and 50 percent female. In some forms, there is a slight deviation from this; in man, for example sex ratio at birth is closer to 52 percent male, and this is true for rabbits cattle and many birds. In domestic chickens, sheep and horses, sex ratios at birth or hatching are usually closer to 49 percent male. The secondary sex ratios or adult sex ratios of many vertebrates show greater diviations, however. In the Alaskan fur seal one adult bull may dominate a breeding group of 30 females.

Wheras the total adult male sex ration is greater than 3 percent the excess adult males are on the periphery of the breeding grounds, they are essentially surplus, and they suffer substa-ntially higher mortality. In most primate populations, adult females out number adult males two or three to one. Apparently males have higher mortality rates throughout life than females have higher mortality rates than males, particularly during the nesting season. In waterfowl and gallinaceous birds, for example male often outnumber females. The sex ratio of blue-winged seal (Querquedula discors) in fall and winter flocks averaged 59 percent males in a sample of 5,090 birds and in the early summer breeding population of mallards (Anas platyrhynchos) and pintail (Anas acuta) in Manitoba, males may outnumber females 4 or 5 to 1.

Age ratios represent another important element of population structure, and one that is also particularly valuable in analyzing populations dynamics Figure2 shows age structure of populations in terms of three basis types: (A) a declining population, with a low percentage of young in the population (B) a stable population with a large percentage of young than adults, and (C) an increasing population, with a very large percentage of young. The exact shape and proportions of these age pyramids are a function of natality, mortality and population turnover. Some populations characteristically have tremendous production of sex cells and youngs with massive morality of young.

The Pacific herring, for example produces 8,000 eggs per famale per season, of which 95 percent hatch but only 0.1 percent survive to maturity. The female shad lays from 30,000 to 100,000 eggs per female per season, and less than 0.1 percent grow into mature fishes. Populations as such would have age pyramids with very broaad bases with tiny segments ascending into adult levels. At the other extreme, elephants and whales usually produce only one young per female every few years, and they would have age pyramids with relatively narrow bases and expanded adult proportions. That is, the majority of animals in the population would be adults.

Rhesus populations in India show different age structures in different habitats. In roadside habitats and villages, where juvenile rhesus monkey (1 to 3 year of age, postweaned but preadult) are trapped for export for use in biomedical research and pharmaceutical production throughout the world, there is a conspicuous juvenile "gap" that is, an unnatural shortage of these individuals. These populations have been declining throughout the 1950's and 1960's at the rate of approximately 5 percent per year, as India has exported any where from 40,000 to over 100,000 rhesus monkey per year, primarily juveniles.

In temples where rhesus monkeys are protected to a certain extent for religious reasons, and in forests where they can escape trapping more successfully age structure of an island population of rhesus monkeys on Cayo Santiago Island off the eastern coast of Puerto Rico is also shown. This population has been fed and protected by a governmental research program of the National Institutes of Health, and it shows a broad base of infants and juveniles. This population was increasing at the rate of 16 percent per year in the early 1960's and gave a good indication of the natural tendency of primate populations to increase if protected in a satisfactory habitat.

Life tables represent tabular data on age structures of populations, and they also provide information on mortality rates and longvity patterns. The essential data of a life table may be of two basic types: (1) census data on a population with accurate counts of the numbers of individuals in each age category, (2)mortality data on the number of individuals of each age group dying in a given period of years. From such data, surviviorship and longevity can be calculated Insurance companies use life table statistics on human populations to estimate the probability or death at each age group, and from these data they establish rates for life insurance. Life tables provide ecologists with essential information for the management of animal population. They pinpoint mortality patterns, indicating when deaths occur and how many. If one finds abnormally high mortality occurring in infancy, adolescence, or early adulthood, it may suggest specific management procedures to alleviate this mortality.

Excessive infant mortality, for example, may be associated with inadequate food for young animals, or inadequate cover by which they can escape predators, or improper patterns of parental care due to behavioral disturbances within the adult population. Excessive adult mortality might indicate shortage of adult food supplies, infectious diseases or parasitic burdens which become more severe in adulthood, or possibly excessive crowding during the reproductive period so that sexual fighting becomes violent. Thus, life data provide important clues for more detailed study. They seldom give the final answers on why or how certain events occur, but they form the essential starting point of what is happening in a population.

We have noted that populations may be studied in both anatomical and physiological terms. That is, we may analyze the structure of a population in terms of numbers, density, spatial distribution, sex and age ratios, and other descriptive qualities. Similarly, we may analyze patterns, movements, and other dynamic attributes. The ultimate objectives are to understand the organization of populations and their vital processes. As we see more and more populations getting out of balance, and going toward extremes of great abundance or great rarity and extinction, such understanding has now become critical to human affairs and the quality of life on earth.

Population Theories

The question of population has occupied the attention of state men and philosophers since ancient times. But it is only recently that attempts have been made to investigate systematically the cause of population change and the specific ways in which population dynamics influence human welfare. Plato and Aristotle were interested in the question of population size in the context of a city-state. To them the optimum size of the population was that is with which man's potentialities could be fully developed and his "highest good" realized. Thus was possible when the population range was enough to be economically self-sufficient and capable of defending itself, but not too large for constitutional government.

Plato specified 5,040 as the number of citizens "most likely to be useful to all cities". European authors on population during the early modern and the medieval periods favoured a growing population. The discovery of the New World, the increase of commerce between Europe and Asia, and the rise of national state brought about some changes in the terms of discussion of population questions, but not until the later part of the eighteenth century was there any marked shift in the general attitude favouring large and increasing population.

The mercantilist and cameralist schools of political economy, which flourished in Europe during much of the seventeenth and the eighteenth century, emphasized the economic, political and military advantages of large and growing population and favoured various measures to stimulate population growth. These writers were concerned primarily with the ways and means of increasing the wealth and power of the State. Their aim was not to promote per capita income but to increase the aggregate national income which was viewed as a sources of revenue for the State.

During the last half of the eighteenth century, a number of writers on economic and social questions rejected the mercantilist doctrine and the view that population growth was advantageous and should be actively encouraged by the State. Some writers, specially in England, France and Italy, emphasised the dependence of population upon means of subsistence. They opposed arrangement for poor relief on the ground that it might undermine frugality, make labour immobile, depress productivity and thus increase the pressure of number on means of subsistence. These arguments were turned against such advocates of social reform as Godwin and Condorect, in an effort to show that any benefits from reform would be cancelled by a consequent increase of populations.

At the close of eighteenth century, Thomas Malthus, an English economist, noted this tendency for populations, including human population to increase rapidly. In 1798, he published his famous work, Essay on the principal of population, in which he put forth the idea that populations tend to increase faster than their means of subsistence. In general, he felt that populations tended to increase geometrically or exponentially, whereas their food supplies and means of subsistence tended to increase only arithmetically.

A geometric increase is a series of numbers having a common ratio, such as 2...4...8...16...32...64 in which each succeeding number is twice the former. An arithmetic increase is a series of number having a common difference, such as 2...4...6...8...10...12, in which each number differs from the other by 2. Thus, Malthus thought that populations constantly tended to outstrip their food supplies and were then decimated by poverty, starvation, disease, warfare, or other catastrophes.

Malthus was primarily concerned with human populations, and he felt that moral restraint was the solution to human population pressures. Malthus's theory is thus based on two postulates and one assumption. His postulates are that (i) food is necessary for the existence of man, and (ii) that passion between sexes is necessary, and will remain nearly in its present form. The assumption is that the production of food increases in arithmetical progression while the population increases in geometrical progression.

The power of population is indefinitely greater than the power of the earth to produce subsistence for men. Efforts must, therefore, be made to keep the population from growing beyond the means of subsistence through some powerful checks, else it will lead to "vice and misery." Malthus talked of two kinds of checks to population positive and preventive. He termed preventive check as "prudence" and included in it postponement of marriage and restraint on reproduction. Positive checks he termed as "natural" as "they arise form the situation itself" and included in it wars, pestilence, epidemics, diseases, famines, natural calamities and the like.

Malthus was highly praised and also bitterly criticised for his views. His Essay aroused a storm of controversy which long out-lived Malthus himself, and which made both his followers and his opponents conscious of the need for adequate information on population trends and their influence on social and economic conditions. Thus Malthus was indirectly responsible for developing population censuses a vital statistics. Malthus has been criticised for his postulates, his assumptions and his conclusions. While it is not to be disputed that food is necessary for man, but so is water clothes, shelter, and such other prime necessities of life.

There is enough historical evidence to indicate that the output of food and of other necessities has increased faster than population, and has been brought about without any distinct symptoms of increasing "misery and vice", but rather of their diminution. With improvement in techniques of production increasing returns are being obtained from land and from the production of prime necessities of life. Malthus has proved a false prophet. While he postulated great miseries on account of unprecedented growth of population, it is found that birth rate in most of the advanced countries of the world has declined, and that the rapid growth of population in under-developed countries is basically due to a rapid fall in death rate and not on account, as Malthus feared, of an increase in birth rate.

Marx rejected the Malthusion theory of population. He held that there is no universal law of population, and that the source of "over-population" was not to be found in man's biological power of reproduction, but in prevailing capitalistic mode of production. Over population arose because capital accumulated less rapidly than did the supply of labouring population was not only a necessary consequence of capita-listic accumulation, but also a necessary condition for the existence and continuation of the capitalistic form of economic organisation.

The realisation that Malthus has over-generalised a particular case lead to a re-thinking of the population question. Writers at the beginning of the present century began thinking that population growth is not always undesirable. This led to the development of the optimum theory of population which is associated with the name of Professor Cannan and others.

The theory says that given the natural resources and the technique of production, that population, Cannan envisaged a point of maximum return for all industries taken together, that is the population which will maximise productivity. He also indicated that the optimum size of the population changes as circumstances change and as new techniques of production are introduced. Effort to precisely define the "optimum" led to the refinement of the concept, but at the same time to the theory being criticised as of little practical value. It is considered highly artificial to select "average real income "as the soletest of the optimum.

In fact there can be an optimum size of the population from ecological, political and other points of view. Again, the optimum size is subject to frequent changes on account of changes in techniques of production. The transition theory, development in the 30's of third century and associated with names of Warmn S. Thompson and Frank W. Notestein, explains the relationship between population with and economic development. It says that in developing countries, which are characterised by heavy pressure of population, low productivity heavy dependence on agriculture, rudimentary techniques of production, under-developing means of transport, and poor sanitary conditions, birth and death rate high.

Death rates are high because of poor diet, primitive sanitation, and the absence of preventive and curative medical facilities. Birth rates are high because of social beliefs and customs which encourage large families and which group up if a high death-rate community is to continue its existence. The birth rate around 40 per one thousand person and the death rate around 35 so that the population does not grow at a fast rate. With economic progress the death rate begins to decline. This is because of improved means of transport, better sewage disposal and improvement in drinking water facilities. But the birth rate stays high so that the difference population tend to grow at the rate of 20-30 per thousand population per year.

Because of the rapid rate at which the population

grows, this period is also known as that of "population explosion". One of the features of economic development is typically increasing urbanisation and in an urban setting children are usually more of the burden and less of an asset. The processed economic change also weakens the force of traditional customs and beliefs.

Females are being to realize that if burdened with a large family they are unlikely to play their due role in society. Consequently, the idea of large family yields place to the small-family-death and the birth rate declines from the high 40 to about 17 per one thousand population. The death rate is also low, about 8 per one thousand, so that the population just maintains itself. The three stages described above are also called the stages of slow population growth, rapid population growth and of stable or declining population growth. The industrially advanced countries of the world have passed through the first-two stages and are current in third stage.

Developing countries of Asia, Africa and Latin America are either in the first stage or entering the second stage. The sequence of events described by the theory can be traced in every region when the economy has evolved from an agrarian to an industrial market oriented economy. But it is not specific to quantity if the degree of decline of the vital rates. The theory, however, contains one significant generalisation, namely, the decline in birth rate occur after a substantial time-lag in comparing with the decline in death rate, and that during this lag population increases very rapidly.

For instance, the "area of European settlement" increased its population six fold during the period 1750 and 1950. The population more than doubled in the century from 1750 to 1850 and nearly tripled in the period 1850 to 1950. The theory is of considerable significance to under developed countires of Asia, Africa and Latin America. Their birth rate is high and the field of public health, it has been possible for the death rate to be substantially reduced with marked changes in the economy or in the birth rate. Consequently, their population is growing at a rate which doubles itself in twenty to twenty-five years. This is tending to retard economic progress. There is therefore an urgent need for bringing about a substantial decline in birth rates if economic development is to become a continuing process.

The first major scientific challenge to Malthusian doctrine as a pattern of population growth appeared in the 1830's in the work of Pierre Verhulst. In 1839 Verhulst proposed that population normally grow in a much more orderly fashion than that proposed by Malthus, describing in their growth a curve of S-shaped proportion. This became known as the Logistic Theory of population growth. The logistic theory asserted that population have a slow initial growth rate, which increases exponentially unitl it reaches a maximum, and then becomes progressively less as the population approaches an upper limit of its growth. The upper limit is approached gradually and in an orderly, predictable manner.

The resulting curve is the familiar S-shaped of population growth. It should be noted that the logistic curve and the Malthusian or irrupting curve do not differ in the early stages of population growth—both show a slow start followed by a period of exponential or geometric growth but they do differ fundamentally in the upper or controlling stages of growth. The Malthusian curve is characterized by an erratic, often catastrophic, pattern of limiting growth, whereas the logistic curve characterized by a smooth, orderly and gradual pattern.

Following the work of Verhulst, the logistic theory lay dormant unto it was independently derived and popularized by Raymond Pearl in the 1920's. Pearl applied the logistic curve to the population growth of yeast, Drosophila and man. This ushered forth a burst of activity resulting in the application of the logistic curve to the population growth of protozoa, water fleas, pond snails, thrips, ants, bees and other organisms by many investigators. With this wealth supporting evidence, the logistic thoery attained wide acceptance. It was proclaimed as a law of population growth by some biologists and was used to predict future population levels of experimental and natural population. Pearl's investigation led him to conclude: "It has been shown in what has preceded that populations of living thing `from the simplest' represented by yeast, to the highest, represented by man, grow in accordance with the logistic curve).

One can now feel more certain that this curve is a first and tolerable close approximation to a real law of growth for human populations.... We can, for example, upon a more adequate scientific basis mere than guess-work, predict future population, or estimate past populations, outside the range of known census counts."

That some biologist accepted Pearl's interpretation rather freely is indicated by statements of Clarke (1954): "Population of a wide variety of organisms, ranging from bacteria to whales, have been found to follow the logistic curve in the growth from.... The growth of man's population follows a similar pattern whether examined in individual regions or in the world as a whole. "Clarke presented a graph of population growth in the United States fitted with a logistic curve, and observed an upper symptote of 184 million in year 2100 A.D.

Many other biologists and mathematicians, however, severely criticized the significance of the logistic theory. In fact, the theory met stern opposition since its first expression could be made to fit a wide variety of curves, often of very different nature. Gray, Hongben, and Wilson and Puffer registered doubt as to the universal validity of ligistic theory and commented on its inadequacies as a description. Feller pointed out that the actual fit of the logistic curve to much of the published data was poor and better agreement could be obtained with other formulas.

Andrewartha and Birch and sang accurately pointed to limitations of the logistic equation. Sang in careful reexamination of population growth in Drosophila, emphasized that, "The data we have summarized show that only in very exceptional circumstances would logistic growth occur in a Drosophila culture... The ecological situation is too complex to be adequately described as the Pear-Verhulst law, "Sang continued to discuss general aspect of the logistic theory by stating, "This acceptance of a plausible formula, in spite of its inaccuracy, may be justified; provided this attitude does not inhibit future research, particularly in to fundamental conception. In the case reviewed, it seems that wide acceptance of the logistic law has led to just this kind of inhibition of further work, and that many interesting ecological and physiological processes have been ignored as a result.

The study of vertebrate population in both natural and confined circumstar.ces has produced some population growth curves deviating considerably for logistic growth and tending to ward errupting or Malthusian growth. The sheep population of southern Australia showeed a pattern of liogistic-like growth from 1840 to 1890, but its upper level was reached abruptly and was followed by substantial mortality. The deer of Kaibab National Forest in Arizona grew from an estimated population of 4,000 animals in 1918 after a period of extensive predator control to an estirnated population of 1,00,000 in 1924 but in the winter of 1924, they died by the thousands after the population had destroyed its food supply. Their growth and decline described a classic Malthusian curve.

The pheasant population of Protection Island, Washington, showed a phenomenal increase from 8 birds on the 397 acre island in 1937 to nearly 2,000 birds by 1942. In one year the population showed an increase of 3000 percent. Unfortunately the field research on this population was terminated before natural control occurred, but the population seemed to be describing a typical irrupting population curve. Several studies of seminatural populations of wild house mice have also shown irrupting population curves.

Seminatural populations have been maintained in confined, but free-ranging conditions with a natural habit for this species, except for the element of confinement. Various factors such as space, food supply, resting cover and the structure of the physical environment have been experimentally controlled to elucidate their roles in population growth and limitation Work of this type at the University of Wiconsin and the Institute of Ecology in Warsaw Poland, has shown that population growth in *house mice* is highly variable, and often dependent upon patterns of sicial behaviour within the population. For example, populations in identical environments showed widely variable patterns of population growth. The variability was related to social behaviour within the population in terms of aggressive behavior, dominance hierarchies, sexual behaviour and parental care.

All populations become controlled even though abundant food and nesting space was still available when aggressive behavior increased to an abnormal degree, when the nesting territories of females were no longer properly maintained, and when parental care of the nestlings become aberrant. Infant mortality increased through nest desertion and cannibalism. The densities at which these behavioral pathologies occurred were highly variable between populations, and they were function of the crowding tolerance and behavioral stability of individuals. This complicated interplay of ecology and social behavior was not adequately accounted for by the logistic theory and the resulting population curves were totally individualistic and unpredictable.

A fairly complete evaluation of logistic theory has been presented by Andrewartha and Birch. They have analyzed the inherent assumptions of the logistic equation and have reviewed many studies in the light of these assumptions. The analysis has led them to conclude that the logistic curve provides an inperfect, yet useful, description of the population growth of organisms with simple life histories (such as yeast and Paramecium); but for animals with complex life histories (such as insects and vertebrates) the population growth usually does not conform closely to the logistic curve except for small parts of the data. The concluding attitude of Andrewartha and Birch is well experessed in the following. "despite its theoretical limitations, the logistc curve remains a useful tool for the ecologist. Because of its limitations, too much reliance should not be placed on it in particular cases until it has been verified empirically for each case. "Hence, it appears fallacious to utilize the logistic equation of a prognostic purpose in any population without experimental precedence.

The evidence is clear, therefore, to support the early contention of Wilson and Puffer that the logistic formula should not be considered a law of population growth to such an extent that it permits extrapolation of the curve of forecasting purposes or the interpretation that the constants of the equation are the constants for nature. Rather, it does provide a convenient description of growth in certain well-studied populations. The major question in human terms is, of course, what pattern of population growth characterizes man? From the early seventeenth century to the present time, the world's population has displayed exponential growth.

There seems to be widespread faith that human populations shall ultimately reveal a logistic pattern, but certainly all attempts to predict human population levels by means of the logistic model have been very far from the mark. Pearl predicted a world population of 2.6 million by the year 2100. This population level was exceeded many years ago, and economic planning based on Pearl's prediction would indeed have been disastrous. Although we continue to hope that human population throughout the world will achieve a pattern of graduallogistic balance through orderly and intentional process, there is no sound scientific evidence that this will in fact occur.

The possible of catastrophic mortality through war, disease and starvation are still very much with us. So also are other patterns of mortality which are not necessarily catastrophic in their total effect, but are clearly related to density, crowding and modern population pressures. In this category are automobile accident mortalities, which are responsible for 50,000 deaths per year in the United States alone, various stress-related diseases, including coronary artery disease and high blood pressure, and various pollution-related diseases including emphysema and chronic bronchitis. If deaths attributable to these causes continue to show present trends, these mortality factors will begin to play a more significant role in population regulation, especially in technologically advanced countries.

Different patterns of population growth may be exhibited by populations recovering from major depletions, or by populations invading new inhabitants. Since most populations are already established, however, and are not growing de novo, they are instead showing varying degrees of fluctuations. Our familiar concept of the balance of nature implies stability but the fact is that most animal populations under go continual change. Natural populations often exhibit a dynamic resiliency, rising and falling in response to many factors. In one sense, a pattern of population fluctuation such as that normally exhibited by most species, may be considered continuous series of growth & decline sequeuces. Thus study of population fluctuation becomes closely allied with the study of population growth and natural control.

Seasonal fluctuations

At the outset it is convenient to separate population fluctuations which are seasonal (i.e., obviously related to seasonal weather), from those which are nonseasonal latitudes, most animal pupulations have breeding seasons so that population growth typically occurs in the spring and summer seasons.

Most arthropods and vertebrates begin producing young in the spring and summer seasons, so that these seasons are characterized by population growth, and they cease the production of young in laten autumn and winter. In aquatic ecosystems, many populations also show marked seasonal fluctuations. Both phytoplankton (miscoroscopic algae), and zooplankton (protozoa rotifiers, small crustacea, etc.) usually show spring and fall increases in population. These sharp increases, often 10 to 20 fold, are known as "pulses". They are sometimes correlated with temperature changes or with turnovers in the water state which recirculate nutrients.

In the tropical regions of the world, where there are not sharpiy defined seasons on a temperature basis, reproductive seasonality still occurs in many plants and animals, and it is often related to the cycle of wet and dry seasons. Many tropical insects, for example, have sharp peaks of abundance related to the monsoon seasons. Even vertebrate animals, including primates, have been found in nature to have definite periods of breeding and birth seasons, often correlated with the monsoons. In both tropic and temperate zones an important principle seems to be that the young appear at the most favourable time of the year in term of food and climate. Thus young deer and antelope are born when insect food is most readily available.

These are exceptions to this general principle; for example, young rhesus monkeys are born in the midst of the hot dry season in northern India, when temperatures are excessively high (often over 100 degree F), water is extremely scarce, and plant growth is virtually nil. In this case, however, the infant feeds exclusively on the mother's milk for two or three months until the monsoon season starts. It is possible that natural selection has acted in such a way as to place breeding, conception and the first trimester of pregnancy at the most favorable time of the year, namely, the immediate postmonsoon month when food is most abundant. In other words, the events of conception and early embryonic growth may have been more critical in the life cycle of the rhesus monkey in terms of evironmental influences than the period of early infancy which is well protected by the mother and the social group. This is speculation and not established fact at this time. It also fits with the observation, however, that the monsoon season with its luxuriant growth again arrives when the infant is two or three months old and is relying less on maternal milk and starts taking its own food.

Population fluctuations which are relatively independent of seasons are of two general types: (1) random, or (2) cyclic. Cyclic fluctuations are those which have a definite periodicity, or rhythmicity; that is, the population reaches a peak at fairly regular intervals. In vertebrate populations, certain species reach population peaks every 4 years, while other have a 10-year cycle. These will be discussed later as specialized cases of more generalized random fluctuations.

Ramdom fluctuations

Ramdom fluctuations may be minor perturbations of fairly stable population, producting relatively flat curves, or they may be major changes in abundance, producing erratic unstable curves. Examples of stable populations are increasingly difficult to find in the modern world. They were probably characteristic of many organisms in undisturbed ecosystems; that is, ecosystems in balance. They were also probably characteristic of complex systems, such as forests and particularly tropical forests where great species diversity produced a complicated web of natural check and balances on each species.

Within such a system most vertebrate populations were probably stable and exhibited only their typical seasonal fluctuations. Probably some of the best modern examples of stable populations may be found among the birds and larger mammals. In birds where considerable species diversity occurs and fairly extensive food resources are available, considerable stability in population size may be present. John Gibb reviewed a number of bird population studies several years ago and noted: "Notwithstanding the considerable annual fluctuation is bird populations that have been recorded in almost every species investigated, it is their comparative stability over long periods that is their most remarkable attribute. "Thus, although wide variations is number occur form year to year in many bird population, these population often show a tendency to return to typical levels.

An example of a relatively stable vertebrate population is afforded by a census study of rhesus monkey populations (Macaca mulatta) in northern India. With complete protection, good health and abundant food supplies, rhesus populations can increase in experimental colonies at the rate of 15 per cent per year. But in India, where monkey experience limited food supplies, varying degrees of protection from the human population, and a considerable number of infections diseases, populations do not show this pattern of increase. In general, throughout India as in most of the world, primate population are declining due to habitat deterioration (cutting of forest, overgrazing of savannah lands, etc.) human competition (where native people shoot monkeys for food), and high rates of trapping for commercial trade. As an example of the latter, in the late 1950's India exported more than 100,000 monkeys per year for use in biomedical research and pharmaceutical production, though the current figure is only about onehalf of this. These monkeys were essential in the development of polio vaccine, measeles vaccine, and many types of medical and psychological research.

The long term population trend of many animals, especially wild vertebrate animals, is down ward. Since 1600 A.D., 120 species of birds and mammals have become extinct, a number several times greater than the natural rate of extinction. At the present time, approximately one vertebrate animal become extinct on the face of the earth every year, and within the next 30 to 50 years, 100 more species will probably be lost. The International Union of the Conservation of Nature periodically publishes a list of rare and endangered species (the publication is known as the Red Data Book), and it now lists nearly 300 species of birds and 275 species of mammals as endangered. The peregrine falcon (Falco peregrinus), ospery (Pandion haliaetus), and bald eagle (Haliaectus leuocephalus) are presently among the vanishing animals of North America. Others have come close to extinction, but were saved by intensive conservation movement at the last moment.

Throughout the world, the basic causes of extinction are habitat deterioration, poaching, poisoning, and direct or indirect competition with man. From the ecological standpoint, some extinction is an inevitable evolutionary process, but when it becomes excessive as in the last century, it is a serious disruption to biotic communities and ecosystems. Each extinction lessens species diversity, and thereby reduces ecosystem stability. Each estinction also represents the irreplaceable loss of unique biological material.

Some vertebrate animals have increased drastically within the last century, but these are usually domestic animals such as dogs, cats, cattle, goats and so forth, or wild animals that have learned to become commensal with man; that is, animals which have successfully adapted and even exploited the human environment. This includes the domestic rats and mice such as the Norway rat, the house mouse the Asian bandicoot rat (Bandicota bengalensis),the Asian house shrew (Suncus murina) to starling, the house sparrow and the pigeon (Columba livea.) Even the herring gull (Larus argentatus) is adapting to man's urban environment successfully by feeding on garbage dumps, and gull population on both United States coasts have increased in recent years.

Apart from those species which seem to be headed towards a permanent decline and eventual extinction, and those which thrive on man-made environments, the fact remains that most animal population fluctuate substantially from year to year. It is often a matter of common knowledge among farmers and sportsmen that "this is a good year for pheasants,"or "there were rabbits all over the place last year, but not so many this year," or "this is the world year for mosquitoes we've ever had," and so forth. These are all popular expressions of population fluctuations.

Scientifically, these fluctuations are not often recorded in a systematic and continuous fashion, but some data do exist to document the types of fluctuations which are common in animal populations. In trying to manage animal populations, we tend to do the most obvious thing, which would logically seem to produce results, and often does, but in the long run, may be quite detrimental to the total ecological balance of a community. Thus, in controlling insect populations, we use residual poisons such as DDT, but are now discovering after some 30 years, that this has long term side effects which are quite dangerous, and we are also dicovering that many insects have become resistant to it in any case. In controlling the population levels and general health of their prey species.

In enhancing the abundance of desirable species, we have often poured money into artificial breeding and restocking programes of animals such as pheasant, wild turkeys and trout only to find that in the meantime, the habitat has so deteriorated through pollution that the transplants have little or no change of success, and the recovery rate per transplant is almost nil. This is not intended as a blanket condemnation of all of our control and management programs, but by and large the balance sheet is not favourable, mainly because we have based our management on trial and error rather than on sound basic knowledge of the ecological factors involved in population growth and regulation.

The most diamatic population fluctuations are those which are cyclic: that is, in which the population reaches peak numbers at fairly regular intervals. Some animal populations, such as those of the snowshoe hare (Lepus americana) have periodicity of abundance of 9 to 10 years. Other animal populations have a cyclical periodicity of 3 to 4 years. These include the Norway lemming of Scandinavia (Lemmus lemmus), the field mouse (Microtus agrestis) of Europe, and some of the predators of these two notably the colored fox (Vulpes fulva) and the snowy owl (Nyctea scandiaca). Not all populations of these animals always fluctuate cyclically. When they appear to do so, sophisticated statistics may be required for proof of real cyclic periodicity. These animal population cycles have been intriguing and puzzling to biologists for many years.

Considerable research has been centered around trying to·understand the cause of cycles, and numerous theories have been proposed to account for them. The final answers are still not clear, though many biological phenomena have been discovered and analyzed in this search for understanding. In general, two types of factors have been considered as the basic causes of animal population cycles:(1) extrinsic factors; that is, those outside of the population itself, such as climate and weather, cosmic events competition between species, predator-prey relationship's and food supply; or (2) intrinsic factors; that is those within the animals as population increases occur and they become crowded.

Several of these factors can be dismissed rather quickly, since extenive research has failed to show any clear relationship. For example climatic and meteorological changes have not been satisfactorily correlated with regular long-term cycles of population change in any animal. Sometimes population peaks coincide with years of great rainfall, or some other notable aspect of weather, but these occurrences are by no means consistent.

At one time, sunspot cycles were considered to be correlated with population cycles. Sunspot maixma have occurred every 10 to 11 years, and they were correlated with peak population of hares and lynx in the late eighteenth and early ninteenthe century, but in the late eighteeth and early ninteenth century, but in the last 100 years, have been entirely out of phase. Changes of populations of predators such as the lynx, fox and sowy owl are clearly correlated with cycles of their prey. But they cannot be shown to cause the prey cycles, for prey cycles often occur in areas devoid of predators. For example snowshoe hare populations have cycles in areas where no lynx are found.

There has been some attention given to disease and parastism as the cause of long-term cycles. Twenty years ago there was interest in hypoglycemic shock in snowshoe hares. Animals at peak periods of abundance seemed to have low blood sugar levels and go into states of shock and death. This was interpreted as the alarm reaction of Selye's general adaptions syndrome. There has also been considerable work on tuberculosis and toxoplasmosis in wild voles and although high disease rates have occasionally been asociated with die-offs of peak population, these associations have not been consistent.

Many rodent declines have occurred without evidence of an infectious disease as the cause. Most ecologists have not been well trained in pathology, bacteriology and virology, not have scientists in these disciplines extended their work satisfactorily to population ecology. Food supply has often been considered inportant in the decline of animal populations. Lack proposed that food is the main limiting factor in bird populations as it is for deer population in North America, Most field studies of rodents have demonstrated however, that even at peak periods of abundance ample food is present. Chitty studied grass during and after high vole plagues and found ample food. There are exceptions; for instance, during lemming peaks food becomes severely depleted and this is one of the factors that is though to start lemming migrations. Pitelka felt that insufficient attention has been paaid to qualitative study of food.

The estrogents and steroid in growing plants, particularly grasses and legumes, are now being studied. It has been observed in the Orients, for example, that upsurges of rodent populations are often timed with the flowering and sprouting of bamboo. Then there is not only increased supply, but the young growing plants are more nutritive and estrogens are increased. The estrogens have direct influence on fertility and fecundity of the animals feeding on them.

2 Social Behaviour

Territorialism is essentially a social pattern of spatial utilization by which individual or groups have control over certain units of space. Dominance hierarchies are an-ordered systems which determine individual access and priorities to natural resources. Both of these phenomena relate to population ecology intimately because they represent systems of behavioural control which affect the abundance and distribution of animals, their reproductive patterns and their mortality patterns.

A territory has been defined as any area which is defended against other member of the same species. It may be occupied and defended by an individual as in the case of a male stickleback fish (Gasterosteus aculeatus); by a pair as in the case of many birds or by a social group, as in the case of vervet monkeys (Cercopithecus aethiops) and gibbons (Hylobates lar). It may encompass most or all of the occupant's home range including his nesting and foraging areas, as in primate territories; it may include only a limited area around a nest site of feeding site, as in many birds; or it may involve only a very small plot of ground on which mating occurs, as in the Uganda kob (Adenota kob), an African antelope. Hence, territorialism is a highly variable and complex phenomenon, and involves a wide spectrum of behaviour patterns. In all cases, however the essential feature is that individual animals, or groups of animals, have controlling ownership over a given piece of property. Within this property they have the right of utilization over the space and the resources contained therein. Territories are established and maintained by a great variety of behaviour patterns and communicative displays. Initially, the establishment of a territory may involve active aggressive behaviour and fighting.

The territorial combats of siamese fighting fish, jungle fowl, elephant seals and wild antelope are well known. Those combats are often vigorous and are true tests of strength, injuries are common, and in rare cases, death may occur. Most frequently, however, the vanquished animal merely retreats and vacates the territory. Once established territories are most often maintained by communicative signals and displays. These signals may be visual, auditory or olfactory in nature. Most fish have visual territorial displays which consist of spreading the fins and brightening the colors. Some fish have active sound production in territorial calls, as for example the Atlantic toadfish (Opsanus tau).

Most birds which are territorial have a combination of vocalizations and visual displays. The male song bird of many species perches in a conspicuous spot and repeatedly gives forth his song, which serves more often as territorial threat than as a mating attraction. Mammals typically use a combination of auditory and olfactory signals as communicative displays in the maintenance of territory. Some of the arboreal primates, such as the gibbon and the howler monkey (Alouatta palliata) have an advanced vocal pattern which conveys information about the position of the group and indicates that the territory is occupied. Many carnivores, such as the wolf, lion and tiger (Felis tigris) have territorial calls which can be heard for several miles, and they also have olfactory signals in scent glands and urine which are used to mark the boundaries of territories.

The scent signals may persist for days or even weeks, and thus serve as more stable cues than serve as more stable cues than vocalizations which fade rapidly. Territorialism is common among many higher invertebrates (crustacea, insects, spiders, etc.) and throughout all vertebrate classes. Territorialism can be found in fish amphibia, reptiles, birds and mammals, which seem to operate on much the same principles and have basically the same functions. Although this is true, it must not obscure the fact that there are many animals which do not show territorialism.

Many species do not stake out specific territorial claims and do not actively defend given tracts of habitat. Often one can find closely related species which are territorial in one case, and nonterritorial in another. In ungulates, for example, the Indian blackbuck (Antilope cervicapra) is territorial, whereas the Indian swamp deer (Cervus duvacelli) is not. In primates, the gibbon and the howler monkey are clearly territorial, whereas the rhesus monkey (Macaca mulatta) and the chimpanzee (Pan troglodytes) are not. Even within the same species there are circumstances in which the individuals of some populations are territorial distributed and other circumstance in which they are not. This is true in ungulates for example, many of which are territorial during the rut or mating season, but the same individuals are non-territorial at other times of the year.

Apart from this obvious correlation with breeding, however, there is evidence in some species that territorialism is related to the population density and/or environmental circumstances. In the Japanese salmon (Plecoglossus altiveli), individual fish are territorial at low population density, but as density increases, individuals give up their territories and form a school. Wild house mice (Mus musculus) show similar behaviour with territorialism at low population densities, and mass group behaviour at high population densities. This variability is related to environmental factors and differences in social behaviour of key individuals. In view of these facts, it does not seem warranted to consider territorialism an instinctive pattern of behaviour. Territorialism is certainly common and wide spread, but by no means universal within limited animal groups or even within a single species. It seems more appropriate to consider territorial behaviour a flexible ecological adaptation—a social and behavioural means of utilizing space.

Dominance Hierarchies

Whereas territories function to allot resources on a spatial basis, dominance hierarchies allot resources on an individual priority basis within the same physical space. Hierarchical systems have been observed in all vertebrate classes and in many invertebrates. Dominant animals have priority access to food, mates, nest sites and resting locations.

Dominance is often manifested between animals by single physical displacement by the displacement of one individual by another at a feeding or resting site. Occasionally it involves displays or threats, and rarely it may involve direct fighting. The latter is more common in initial encounters in which the dominance rank or social status of each individual is not yet determined, but once established the hierarchy is then usually established by display or by social memory.

Once a rhesus monkey has established high rank, it does not have to constantly display this rank in a stable social group. The high rank of some individuals and low rank of other become an accepted social norm within the group. Only when the situation becomes unstable through death, emigration or invasion does there have to be an avert re-establishment of rank within a dominance hierarchy. So far as we have seen territorialism and dominance hierarchies as two types social systems which play a regulatory role in population dynamics, the first by regulating of space second by controlling individual priorities within a common space.

In some animal species and populations, territorialism and dominance hierarchies apparently serve as sensitive feedback mechanisms which influence dispersal, mortality reproductive success and population size. In other species and circumstances, territorialism and hierarchical structures are often subject to breakdown and collapse. Ecological consequences are then profound and extensive. In many rodent populations, excessive population growth and crowding is often associated with the following phenomena: a breakdown of stable territoria-lism and dominance structure, reduced fertility as evidence by lowered pregnancy rates, high prenatal loss and low infant survival, poor nest construction and poor maintenance, nest fouling and pollution, high aggression and wounding deteriorating health in the majority of the population, abnormal aggregations of individuals in the 'behavioural sink' pattern and finally population decline.

The detailed study to territorilism and dominance hierachies is really in the realm of ethology and animal behaviour, but the functions and consequences of these behavioural systems have a great deal of ecological significance. It has been abundantly demonstrated in many territorial animals that the holders of territory have the greatest breeding success. In fact, in some animals, non-territorial animals usually fail to breed. Similarly in those social groups characterized by dominance hierarchies the high ranking animals have the greatest breeding success and the low ranking members have less success or fail entirely in breeding efforts. In chicken flocks, low ranking roosters show frequent counting behaviour, sometimes more than dominant rooster, but they have relatively little success in completing copulations. It thus seems clear that the social systems of territorialism and dominance hierarchies have a limiting function in population dynamics; that is, they limit the breeding to those individuals which are ecologically and behaviorally most successful in establishing territory or high rank.

Theoretically, as population densities increase, and space becomes more limited, a smaller proportion of individuals within the population is able to achieve territories or high rank, and thus total reproduction is reduced in relation to population size. Ideally, these behaviors are density-dependent population limiting factors; that is, the higher the density, the greater the limiting effect on the population. The role of social behavior in population regulations has been discussed at great length by Wynne-Edwards. Wynne-Edwards proposed that the basic long-term population level of many animals is established by food and resource availability, but that social behaviour establishes a more immediate regulating influence on population size and density. This influence on social behaviour, he felt, comes into play before critical population levels are achieved; that is, before starvation and/or damaging violent aggression are conspicuous.

Thus, according to Wynne-Edwards most animals have intrinsic, self-regulatory population mechanisms which operate via social and behavioural feedback systems. These social and behavioral feedbacks are usually ritualized displays which relate to territorialism and dominance hierarchies, and thus convey information about population density. Therefore, regulatory feedback is provided according to the intensity of social competition. In other words, ritualized displays, or conventionalized rivalries, regulated breeding success and thus population numbers. Wynne-Edwards provides numerous examples of this hypothesis, including the complex display patterns of gallinaceous birds, (prairie chickens, sharp-tailed grouse, peacocks, bower birds et al.) waterfowl and many song birds. He also considers the vocal choruses of amphibians, the schooling of fish, and the flocking and birds and mammals as examples of complex social behaviour which serve the function of conveying information about population density and social competition.

Many aspects of Wynne-Edwards' theory are controversial especially when he theorized that most social behaviour has actually evolved as a means of controlling population levels. Wynne-Edwards proposed, for example, that territorialism has arisen as a social means of preventing over population. This theory has met substantial criticism from other ecologists and students of behaviour. These authors felt that there is little reliable evidence to support this hypothesis even though it is logically very attractive. Brown pointed out that, "Population regulation is never completely under the control of the species by itself, but depends in a complex way on interactions between members of the ecological community." Thus, Brown felt one must consider the interaction of all factors, including predation, disease, food supply, interspecific dynamics of any given species.

Brown found satisfactory evidence that territorialism tends to space individuals over available habitat, that it increases emigration and morality rates in individuals unsuccessful in finding territories, and that it has arisen in response to individual selection resulting from aggressive competition. The more detailed aspects of this controversy are out side the scope of introductory ecology. On the more basic parts of the hypotheses—namely, that social behaviour provides some regulatory feedback on population ecology—there is certainly no question that this is true in most animal populations. Even this concession must not obscure the fact, however, that some animals have such poorly developed and/or variable responses to increased density and crowding that they can readily breed themselves into starvation or conditions of violent aggression and extreme social strife.

White-tailed deer populations, for example have frequently increased to the point of extensive starvation. Lemming and other rodent populations have also increased to the point of mass behavioral strife where social mortality becomes extensive. Quite probably, some human populations are also in this category, providing example of populations in which social feedback limiting population growth often does not into operation until the point of starvation arrives (as in China and India), or until the point of violent and destruction aggression breaks forth (as in many wards in which population pressure has had a causative influence) or until social morality becomes limiting (as in countries where abortion is a significant form of population regulation). It is scientifically important to consider animal population in which environmental crises and/or behavioural breakdowns occur.

In terms of human value judgments these seem to be cases of population regulation by wasteful and sometimes tragic means. In evolutionary terms, however, many of the populations showing these phenomena are of highly exploitative natures, meet new environmentral situations had expand into available opportunities. Perhaps their lack of frigid, stereotyped behaviour and ecology, and their lack of precisely tuned population feedback systems, enables them to seize each new ecologic opportunity to the fullest extent. Some of the best examples of such species are the domestic rodents especially the wild.house mouse, the wild Norway rat (Pattus norvegious) and the Asian bandicoot rat (Bandicota bengalensis). Numerous ecological studies over the last 25 years have shown these species to have the following population characteristics:

1. They have tremendous reproductive potential and under certain environmental circumstances they often show invasive and eruptive population behaviour.
2. Populations can readily exceed the carrying capacity of the environment and this can result in sudden population decline.
3. Populations fluctuate widely, natural control is often abrupt and "ecologically wasteful", in terms of individual health and welfare.
4. Population regulation often involves social disruption and destructive behaviour.

These characteristics have been observed frequently in both natural and artificial populations of rodents over the past 40 years.

The natural social behaviour of the Norway rat and wild house mouse includes both territorial and hierarchical behaviour. Females tend to the territorial around the nest site so that they normally exclude adult males and strange females. Often two of three females nest together in a communal agreement. Males are also territorial if population density is not excessively high, and high ranking males defend a system of runway and the females living therein.

The size and patterning of the territories depend upon the rank of the male, the density and social structure of the population and the nature of the habitat.

In a stable habitat both population levels and social structures of domestic rats might remain quite stable over a period of time. But the habitats of domestic rodents are characteristically unstable with changing amounts of food (around farm, for example where harvests and food stores come and go), changing mortality patterns (as control compaigns) or disease outbreaks wax and wane) or changing cover conditions(as trash availability rises and falls) so that the populations are often experiencing drastically changing resource conditions. Under these circumstances, the populations respond rapidly with surges or declines of growth. Under periods of rapid population growth leading to high densities, a variety of phenomena have been observed.

There is often rapidly increasing social contact resulting in an increase in aggressive interactions. This frequently lead to a breakdown of territorial boundaries, increased crowding of the nest sites and sharply increased mortality of young nests become trampled and destroyed. Females often desert the nests or actually kill the young. Wounds become frequent, especially an adult males forming sites for mites to invade the skin, so that acarine dermatits may achieve epidemic proportions.

With a breakdown of territorial behaviour, bizarre aggregations developed so that clusters of individuals all in poor physical states appeared. Calhoun referred to such aggregations as "behavioral sinks". They are self-stimulating aggregations of individuals often collectively showing abnormal behaviour patterns. Some individuals become inactive and socially withdrawn. As these conditions progress only limited portions of the environment are utilized but these become excessively polluted.

The stage is set for both high social mortality and high disease mortality and population often collapses as rapidly as it built up. After the crash if the environment recovers the surviving individuals recuperate and population growth may begin again. The detailed of this theme differ considerably in different studies and different species.

In some circumstances, there may be a cessation of production so that no young are born into the crowded population, as occurred in house mouse populations studied by Crowcroft and Row. In other circumstances reproduction may continue, but there may be massive infant mortality of all young born. In still other cases, crowding may stimulate excessive migration and disposal as in lemmings and these mass migrations are often accompanied or determinated by high mortality. In all cases, a similar syndrome prevails. There is usually a drastic change in social organization and behaviour which most often have a major influence on reproduction and mortality. In domestic rodents these changes characteristically involve a breakdown in social stability and a disruption of normal territorial, reproductive and parental behaviour. Frequently parinatal processes are most dramatically affected, so that this pre-and postnatal survival of the young is reduced.

Population Interaction

All population of living organism exist in a network of interactions with other population. Many of these interactions are subtle and complex than direct food relationships. Others are competitive or limiting to the interacting populations. The degree of intimacy in the relationships of the organism inhabiting a particular habitat varies greatly. These association may be either beneficial or harmful to both the patterns or beneficial to one partner and harmful to the other. If the relationship is beneficial either to one or both partners, it is called 'Positive Interaction' or cooperation. This may be interaspecific (Ex. Social life of inspects) or Inter specific (Ex: Commensalism, Mutualism & Symbiosis). The interrelations between individuals that are harmful to one or more of the participants is called Negative interaction. It includes "Parasitism, Predation and Competition".

Commensalims

The term 'Commensalism' was first proposed by Van-Benedan to the animals that share the same food. "Two or more animals that live together and do not enter into any kind of Physiological Union are referred to as `Commensals' and this association is called as `Commensalism! The relation is unequal in that only one individual gets the benefit from the association while the other is neither benefited nor injured. The Chief benefits from this association are (i) Shelter, (ii) Anchorage (iii) Transportation and (iv) The food supply.

Types of commensals:

A commensal which attaches or lives upon the host's body is called as 'ectocommensal'. If a commensal lives within the body of the host in the respiratory or alimentary tract or in any other body cavity which opens to outside, is referred to as endocommensal.

Vorticella, a Protozoan animal is an ecto-commensal which is attached to the legs of the freshwater shrimp (prawn like) by slender stalks. The 'Sucker fish' is another ectocommensal which attaches itself to the body of shark, or sword fish or seaturtle with the help of its sucker. All the ectocommensals are small in size and not burdensome to the host. The harmless protozoans in the intenstinal tract of mammals and micro-organisms in the canal system of sponges are the examples of an ectocommensals!.

An example of commensalism is the remora-shark relationship where the remora fish (Echenesis naucrates) attaches to the skin of the shark by means of a strange sucker disc and transported widely and rapidly by the shark's motive power. The remora also consumes food remnants cast off from the jaws of the shark.

The remora benefits in two ways from this attachment and the shark is relatively unaffected although its swimming speed may be slightly impeded. Many large organisms provide commensal harborage for other organisms externally and internally. Large tropical trees provide habits for numerous commensal plants and animals. Trunks and branches provide attachment sites for epiphytes such as orchids, and recesses between buttress roots provide attachment harborage for bats tree frogs, lizards, insects and many other organisms. Whales provide attachment sites for barnacles, algae, and other sessile marine forms. These plants and animals are not parasitic, that is, they do not extract nutrients from the habitat provided by the host.

Most animals, including man, also contain internal commensals. The human digestive tract is inhabited by several types of bacteria and protozoa such as Entamobea cole, which are neither parasitic not pathogenic, but merely residual in the alimentary canal. Another common type of commensalism occurs when various species of plants and animals use burrows or nests constructed by another species. Termite nests provide ecologic niches for more than 100 species of other animals including ants, aphids, beetles millipedes and isopod crustaceans. Tube-dwelling annelids of the genus Chaetopterus which inhabit the tidal zone of the seashore provide habitats for small crabs of the genus Polyonyx.

The crab benefits by the protection of the tube, and it apparently does no harm to the annelid worm which originally built the tube. Commensalism may be obligatory, in which one organism depends entirely on another for its habitat or it may be facultative in which each organism may be independent but one is enhanced in the presence of the other. An example of obligatory commensalism is the relationship between the algae, Basicladia, and certain freshwater turtles. The algae grows only on the backs of the turtle. Similarly the bivalve mollusc Ostrea grows almost exclusively on the roots of the red mangrove off the coast of Florida. Attached to the shell of the horse-shoe crab, Limulus, are several species of molluscs, barnacles and annelid worms which again are not parasitic but only residential and some of them live only on the Limulus shell.

The oyster crab is a small crustacean which lives only in the mantle cavity is a small crustacean which lives only in the mantle cavity of oysters. These all represent cases in which one species of plant or animals has evolved a dependent relationship does not involve feeding upon the tissues of the host. The possibility exists that this is one evolutionary route toward parasitism and the distinction between obligatory commensalism and parasitism is often difficult to draw. Facultative commensalism is represented by the relationship between the borrowing owl (Speotyto cuniculario) and the prairie dog (Cynomys sp.)

The burrowing owl often nests in prairie dog burrows but is not confined to such burrows. In the north polar regions, the arctic fox Clo pex lagopus) feeds in the winter time on the remains of seals killed by polar bears (Thalarctos maritimus) though it may have other sources of food as well. Probably the majority of commensal relationship are facultative. It is especially interesting to consider animal populations which form facultative commensal relationships with men. Among vertebrates perhaps the best examples are domestic rodents such as house mice and Norway rats which have increased in abundance and distribution by their association with human habitation and agriculture. The house sparrow, pigeon and Asian house crow and Asian house shrew or musk shrew have also greatly increased their population sizes through human associations. Cities, towns and farms provide increased food supplies and nesting sites of these animals. The rhesus monkey in India is one of the few primates which has become a facultative commensal of man. The rhesus can readily survive in forests without man, but it becomes especially abundant around villages and towns of the Gangetic basin where it could live on agricultural produce.

Some more examples among the inverterbrates are as follows:-

(1) The peacrabs are found living in the mouth cavity of bivalves. They steal food collected by the host and do not cause any injury.

(2) A number of micro organism live as commensals in the canal system of sponges.

(3) Polynoid polychaetes are found in the pallial chambers of gastropods and in the ambulacral groves of star fishes.

(4) A host of commensals such as worms, Brittlestars, molluscs, little fishes, prawns-crabs etc. are found among the coral reefs.

(5) Small pelagic fishes live under the umbrellar shelter of large jelly fishes.

(6) Some fishes live in the borrows of worms and crustaceans.

(7) A number of copepeds live within the Ascidiann or Sea anemones.

(8) Barnacles attach on the skin of whales or back of turtles.

In all these associations one of the partners which is generally smaller derives benefits of anchorage, shelter, food and transportation.

Mutualism

Mutualism is an association between two or more animals in which all derive mutual benefit with no harm to any of the partners. "Symbiosis" term also often used to this type of relationship (Properly refers to the intimate associations to two or more dissimilar organisms). Mutualism may be seen between animals and plants and between animals and other animals.

Ist Type : Mutualism between Plants and Animals

This association is important to the reproduction of many kinds of plant species. Plants attract insects by their sweet, nectar or colour. The insects visit plants to get their food. The plants dust them with pollen and the insects carry the pollen from flower to flower.

In the distribution of seeds, plants take the help of animals. Plants provide fruits and seeds pass in their hard coat, undigested through the alimentary canal of the animal and are distributed in its droppings. For instance a species of Alacia grows little pearshaped masses of food for the ants, and hollow thrones (Myrmecolphilous stipules) in which the ants take shelter. In turn the ants defend the plant from the attack of other insects.

2nd Type: Mutualism between Animals and other Animals

Birds display this type of relationship. The cow bird and Bison, the Oxpecker and the antelope, crow and the cattle exhibit mutualistic relationship. In all these associations the birds render considerable service to their hosts by ridding them of ticks and external parasites and also serves as sentinels. In turn the birds receive a constant food supply.

An interesting example is between a 'Crocodile' and the bird (Pluvianus aegyptius). The bird removes leeches and fragments of food in decaying condition of the crocodile. The crocodile opens its jaws and permits the search then finally the bird receives the food from the crocodile. The common starling bird may also perform the same function from the sheep. The association of ostriches and Zaebras is also said to be of mutual benefit.

Birds and Insects

Birds have also developed relationship with insects for mutual benefit. The horn bill of Africa harbours a number of insects in its nest. The female horn bill will be confined to the nest at the time of egg laying and breeding. The male bird feeds her with fruits. Various ants plants (Lice Aphids) are associated as Commensals.

The most common Commensals among marine forms are the Hydroids and Sea anemones found on the shell of Hermit crab. In this association both of them are benefited. The Indian Rock (Mincus inermic services as a moving home to zoophytes small polyps are encrusted on this fish. In this relationship the fishes obtain protection from predators and they pay for it by luring large fishes within the range of tentacles.

The relationship between *Hermit Crab* and *Sea unemone* is very interesting. Here the Crab obtains protection from predatory fish by the shrinking power of sea anemone. The sea anemone also gets an advantage of transportation of feeding grounds and gets food fragments when the crab captures and eats another animal.

Sponges growing on the shell of Hermit Crab protect it by producing bad odour. Here the advantages to the sponge is that it is carried to new waters for its own food. Thus, the mutualistic relationships between plants and animals, and animals and other animals prove to be beneficial mutually and reciprocally to the participants. Like commensalism, mutualism may be obligatory or faculative. An example of obligatory mutualism is provided by the relationship between the flagellate protozoan Trichonympha, and woodeating termites. These species cannot live without each other. The protozoan lives only in the digestive tract of the termite.

The stomach and caecum of the horse contains millions of ciliate protozoa and bacteria which digest cellulose for the horse and provide 20 per cent of its nitrogen requirement per day. These intestinal microorganisms are essential for the normal growth and health of the horse. An elaborate form of obligatory mutualism occurs between some ants (e.g., Cremastogaster lineolata) and aphids or plant lice (e.g. Aphis caliginosa). The ants maintain the aphids in specially constructed nests, and they feed upon a nutrient liquid secreted by the aphids.

The aphids have sometimes been called "cows" because they release the secretion when stroked by the antennae of the ants. Thus the ants receive food from ants. Many other anthropods, including mites, springtails, flies, wasps and beetles, have developed similar complex relationships are based on interspecific communication systems through which the different species jointly utilize food, shelter of other resources. Many mutualistic relationships are facultative; that is, both populations can exist without the other, but both are favoured when living together. Baboons and impalas, for example, can survive well without each other, but each provides a unique protective warning system for the other when they live in association.

Squirrels facilitate the extension and propagation of hickory trees by buying the nuts, but the hickory trees can propogate without squirrels and the squirrels can likewise survive without hickory nuts, if an adequate supply of other food is available. Man has capitalized on mutualism as a biological principle in the development of agriculture. The domestication of plants and animals originated in facultative commensalism or mutualism. At the down of agriculture, 10,000 years ago, man could live without domesticated plants and animals, but as he developed these relationships then evolved into obligatory commensalism and mutualism. Modern man obviously cannot live without the domesticated plants and animals of agriculture, and many of these domesticated forms cannot live without the special husbandry provided by man.

Symbiosis

The term Symbiosis was first coined by DeBray in 1879. It means "living together." "Symbiosis is an association between two or more organisms that live together in close physiological union for the mutual benefit." Here the term Symbiosis refers to the intimate association of two or more dissimilar organism. One of the partners may be known as Symbiont.

Symbiosis offers from commensalism and mutualism in two important respects.

(1) The relation between two partner is much more intimate in Symbiosis than in other relations.

(2) It is mostly obligatory.

Symbiotic associations are known to occur between plants and other plants, animals and plants; and animals and other organisms.

First Type Symbiotic Association among Plants

This type of associations is demonstrated in (i) Fungi and algue which form lichens; (ii) Fungal mycorrhizae and the roots of many flowering plants and (iii) Nitrogen fixing bacteria and the roots of legumes. In all these associations, there is metabolic interdependence. In the lichens, algae produces food and fungi provide mineral and protection. The intimate association between the roots of flowering plants and mycorrhizae results in the formation of a composite fungus-roots structure which differs from the roots without mycorrhizae. Here the fungi require soluble carbohydrates and substances and they get them from root secretions. Here also the association is intimate the benefits are mutual. The importance of the Nitrogen-fixing bacteria (Rhizobia associated with legumes is well known. Legumes posses on their rootlets, a number of swellings, inhabited by the bacteria. Here, they derive carbohydrates from the host and in turn they fix the atmospheric nitrogen in the roots. This "Nitrogen" is used by the plants. This is clearly a beneficial partnership.

Second Type: Symbiotic-Associations Between Plants and Animals

The most notable symbiotic association between plants and animals occur between freshwater coelenterate (Chlorohydra Viridissina) and an algae. In this association the animal obtains Carbohydrates and a supply of Oxygen evolved as a result of Photosynthesis by the algae. The large in turn receive protection and Dioxide for photosynthesis released by the animal during respiration. Hydra and Algae relationship is also one of the good examples. The mutual benefits of hydra an algae are

Chlorachydra (animal)	*Zoochlorellae (Plant)*
i) It gives phosphorus and nitrogen compounds for protein formation.	i) Plants gives Oxygen
ii) It gives Carbon dioxide for the photosynthesis in plants.	ii) Excretory materials.
iii) It gives protection to algae (Plant)	iii) Carbon dioxide (from respiration)

Planeria and an animal algae and a plant symbiotic association is also a good example. Here the symbiotic benefits are same as in coelenterate and algae. Here, the presence of algae is indispensable for the worm and this is obligatory.

Association Between Fungus and Insects

The wood boring beetles have an elaborate apparatus on the head for carrying the Fungus. Others carry Fungus in the midgust to be regurgitated for cultivation. The females of this species have an apparatus connected with evipositor which coat their eggs with fungal spores as they are laid. In all these associations, the Fungus provide food and shelter, and it grows on the excretory material of its host.

Third Type: Symbiotic Association Between Animals (Metazoa) and Microorganisms

Symbiotic associations are also seen between Metazoa and Micro-organisms many Wood-eating-insects especially termites harbour numerous protozoans in their alimentary canal. It has been shown that the gut of termites can live on a nitrogen-free diet so long as protozoa are present. This suggests that they may be capable of nitrogen fixation of the synthesis of protein from carbohydrates. Here the protozoans produce specific enzymes "Cellulose," and "Cellobiase," which are responsible for cellulose digestion. Similar association is seen in the gut of Cockroach, "Cryptocersus." In all these cases the protozoan gets shelter and food.

Mammals and Bacteria

In this association symbolic bacteria are present in the Caecum of rabbits and rumen of cattle. Here, the bacteria utilize the inorganic nitrogen and sulphates for the synthesis of essential amino acids, which can't be synthesized by the animal In turn to mammals, the

Bacteria produce Vitamin B12 (in ruminants). In man also some of the vitamins are produced by the intestinal Bacteria. The Symbolic Bacteria are also present in the gut of the South African Honey Guide." Here, the Bacteria feeds on the Wax of honey comb the bacteria digest the Wax and gets shelter and food from the host.

Predation

In this relationship one animal is killed and eaten by another, the predator so predator is a free living animal. It is usually larger than prey. The herbivores that feed on the plant material completely are called as Plant Predators. There are, many kinds of predation which are not necessarily so selective. The swallow feeding on moths and midges, the top minnow or trout feeding on mosquitoes, and the ant eater feeding on an ant nest is less discriminate in taking individuals.

Many able and healthy individuals are preyed upon. In cases such as these, which involve large numbers, the prey population often shows great productivity which can accommodate this mass mortality. Many fish, insects, and other Invertebrates have tremendous production of eggs and larvae, but only a small percentage of these serve the onslaughts of predation and other forms of loss. For example, the horse shoe crab or Limulus of estuarine and evastal waters deposits its eggs on sandy beaches wehere they are consumed in great numbers by flocks of laughing gulls and other shore-birds. Large number of eggs also wash out to sea and are eaten by fish, whereas others are pushed up on the beach and dry out. Despite this fantastic loss of eggs and young larvae, Limulus continues to survive in large populations and has done so for the last 180 million years. It is one of the oldest animals on earth in terms of its evolutionary history, having outlived trilobites and dinosaurs by millions of years, and it is still incredibly abundant.

There has been considerable interest in predation as a factor in quantitative population control as well as its role in quality control. On one hand, sportsmen have wanted to limit predation, being concerned that predators may take excessive numbers of game birds and wildlife. Some sportsmen's organizations have therefore supported predator control campaigns against hawks, owls, foxes, coyotes, wolves, cougar and other predators in the name of conservation. On the other hand, ecologists have often been interested in promoting predation as a method of biological control of pest animals, particularly insects and rodents, and even as a necessary biological control of desirable animals such as deer, or prevent them from becoming pests.

There are also examples from the entomological literature where predation serves as a significant population control on an animal population. In a classic study of natural population balance in the knapweed galifly (Urophora jaceana), a small dipteran insect, Varley concluded that one of the major controlling factors was the predation of mince upon insect pupae. His data showed that 22 to 43 percent of the fly morality was attributable to this predation. One study of tsetse flies (Glossing spp.) in Africa, found predatory spiders to be a controlling factor on the abundance and distribution of the flies. Insect predators have also been used in applied biological control of agricultural pests in a few cases.

In California, the cottony cushion scale, an hemiptoran insect pest of citrus crops, was effectively controlled by a predatory ladybird bettle imported from Australia. In Hawaii, the sugarcane leaf hopper has been controlled by the capsid bug. Another important aspect of predation as an influence in population ecology and natural selection is the great variety of adaptations which have arisen in response to it. Protective coloration, warning coloration and mimicry fall into this category. Also, the morphological and behaviour, freezing behaviour and many other specialized patterns have developed in various species as adaptations to reduce mortality through predation. One particularly interesting system which has been studied in some detail is the intimate relationship between certain moths and the bats which prey upon them.

The bats locate the moths by emitting ultrasonic pulses and detecting reflecting echoes flying moths, a natural sonar system. The moths have evolved the ability to detect these ultrasonic pulses of the bats, and upon perceiving the approach of a feeding bat, they instantly go into complicatd diversionary flight spirals to avoid the bat. The introduction becomes on serial "dog flight" between bat and moth, with the moth trying to avoid the bat. This ability and behaviour has evolved in selective response to predation in the same way that cryptic coloration has evolved. In both case the advantageous mutations have been selected and propagated.

Parasitism

"Parasitism" is the relation between two individuals, wherein the parasite receives benefit at the expense of the host, therefore, parasitism is form of disoperation. In this association, the parasite receives (i) food (ii) shelter (iii) and protection, at the expense of the host. The host does not derive any benefit from the relationship. A parasite does not ordinarily kill its host atleast not until the parasite completes its reproductive cycle.

Classification

Parasites are classified into different categories depending upon their relation with the host.

1. Temporary Parasites

The animals which spend only a part of their life cycle as parasites are called temporary parasites." For Example;

The glochidium (is a Larva of freshwater mussel) leads a parasitic life. It attaches itself the skin of fish with the help of hooks present on the shell and later it is covered by the skin of fish. It remains burried thus for several weeks and emerges as young animals to lead a free life.

Example 2. In prasitic Hymenoptera, the eggs are deposited into the body of another insect (Caterpillar), where the entire larvae stage is spent the larva feeds on the fat body of the host. Until it attains pupal stage. Finally it emerges as an adult and leads a free living life.

2. Permanent Parasites

Some parasites have continuous association with their hosts, they are called permanent parasites. They spend their entire life as parasites. For example: Trichinea worm, Malarial Parasite etc.,

Permanent prarasites often found among lower animals. They almost always require a succession of hosts or intermediat host.

3. Facultative Parasites

These are parasites (animals) which are parasitic only when there is a change for parasitic life, leading free life otherwise. Eg: Meggocs of some flies, Lamprey etc.,

4. Obligatory Parasite

Some species are forced to lead a Parasitic life. If not, they can't live. This type of parasites are called obligatory parasites. Eg: Filarial Wornm Cestodes etc.,

5. External Parasites

Those which live outside the body of the host are called external parasites. Eg. Flea on birds and mammals, blood sucking lice and fleas, mites, ticks and leeches.

6. Endoparasites

These are parasites which live in the elementary tract

body cavities, various organs, blood or other tissue of the host. The association between 'endoparasites' and their hosts is invariably more inntimate than that between extoparasites and hosts. Tricchomonad flagellates, Opalnid ciliates Sporozoans, Nematodes, Cestodes and some Copepods are some of the endoparasitic forms.

The endo-parasites are further divided into 2 types based upon their exact location inside the most body.

(i) Intracellular Parasites Ex:- Plasmodium (living within the cell)

(ii) Intercellular Parasites Ex:- Tryponasome (living be tween the cells in the intercellualar spaces).

Evolution of Paraitism

The origin of parasitic life in amimals is associated with attempts to secure food for themselves or their offspring. The ancestors of ecto-parasites were clearly free living forms. A free living organism could accidentally have settled on the body of a larger species where coditions were favourable for survival. For example: The biting lice might have evolved from the insects that live beneath the bark of trees. Endoparasites might have evolved from ectoparasites or directly from free-living ancestors.

Modifications and Adaptations

Organs of attachment: The rasping mouth of the Lampern, and the head of tape worm equipped with 'Suckers' and 'Hooks' for attachment to the hosts gutwall are the typical examples. Among parasitic arthropods such as mites other insects claws are well developed used to cling on the external surface of the host.

Body Shape: The intestinal parasites such as the tape worm and Round worm have elongated bodies. The liver fluke have leaf-like bodies and elongated form of blood flukes are good example of this type. These arrangements are useful to suit the environment in which the parasite lives.

Cuticles: Ex: Tape worm. Intestinal parasites face the problem of with-standing the action of the digestive juices of the host. To avoid this these parasities develop thick cuticle, offers resistance against the action of the digestive juices.

Loss of Locomotary Organs: Most of the endoparasites have lost their Locomotary organs, as they are not needed. There is no need in search of food, as they have always plentiful supply of food. So the locomotary organs of the parasites disappear or differentiate into organs of attachment. The fleas or lice (Parasitic Insects) have lost their wings, which live among the hair and feathers.

Digestive System: In the tape worm, which lives in the alimentary canal of the hot, has lost its alimentary canal and absorbs already digested food from the surrounding medium. The mouth parts of the blood sucking insects have developed 'Piercing and Sucking' apparatus.

Loss of Sense Organs: As the internal environment of the host is more or less uniform, there is no use of the sense organs among the endoparasites are provided only with tango receptors.

Complicated Reproduction: The reproductive function in most parasites has undergone considerable complication. It is the goal of all the functions of the organism. The endoparasites are characterized by enormous fertility.

For example (i) the human round worm lay (Ascaris Lumbricoides) eggs at a rate of 1500 a day (ii) the termatodes and cestodes lay eggs continuously (iii) Taenia solium (Tape Worm) lay eggs annually and have been estimated at 80 million (per year). The most remarkable feature of flat Worm parasities is the presence of 'complicated reproduction'. The effect of "Parasitism" on the host is usually harmful. The endoprasites causes effects, more marked than those by extoparasites. The following Pathological effects caused by parasites in their hosts.

(1) The destruction of cells and tissues of the host by the parasite's movement or feeding act. Ex: Entamoeba histolytica-eats the tissue cells of colon and red blood cells of the host, and plasmodium feeds on liver cells and erythrocytes (R.B.C.)

(2) Parasites may cause 'enlargements's and disorders of lymph glands, and liver. Ex: Leishmania.

(3) Parasites may cause 'Malaria', (Plasmodium parasite) and 'Trypanosoma gambiense' diseases due to the secretion of toxins (Poisonous).

(4) The effects of endparasites may also be mechanical. Ex: Ascaris, is a good example when these parasites multiply in greater number they block the intestine, preventing the feree flow of food, which leads to the death of the host.

(5) The most remarkable effects of prasitism on the sexual characters are seen among crustaceans. The presence of sacculina bring spectacular changes in its host (Crab), which may result finally in death (due to the parasitic castration).

Hyperparasites

Sometimes a parasite harbors a number of parasites in its body. They are called 'Hyperparasites'. A typical example is, small mammal ones as 'Mouse' which invariably harbours a no of ectoparasitic fleas, which themselves contain endoparasites (Protozoans) in their alimentary Canal.

Parthenogenic and Non Parthenogenic Parasites:

The parasite which do not cause any injury to the host

are called 'Non parthenogenic parasite'. The parasite which cause disease is known as "Parthenogenic Parasite". In man's digestive system 'Bacteria' are helpful in synthesizing' Vitamins, used to man. Some Bacteria causes numerous diseases as pneumonia, tuberculosis, diptheria, bubonic plague, cholera, syphilis and wooping cough etc.

Viruses causes, small pox, polio, yellow fever etc. Animal-like microbes causes malaria, (Plasmodium) sleeeping sickness (Trypanosona gambiense) and dysentery (Entamoeba histolytica). Parasitism is virtually universal in all plants and animals in vertebrate animals, endoparasites are found within many of the major organ systems of the body, most commonly in the digestive, circulatory, respiratory and urogenitial systems. Ectoparasites occur on or within the skin and its appendages, such as hair scales. Throughout most of the world, man is beset with many parasitic organisms including intestinal worms (tape-worms, flukes, roundworms, etc.), intestinal protozoa (amoeba, ciliates and flagelates), parasites of the blood (microfilaria, malaria protozoa, etc.), and a variety of extoparasites (Mice, mites, ticks, mosquitoes, etc.).

Many infectious microorganisms including bacteria and viruses are also parasites in the sense that they extract nutrients from the host. Some of these are pathogenic, that is, they impart the health and normal functioning of the host, as for example, the bacterial organisms of the genus shigella which cause bacillary dysentery. Others, such as some intestinal flagellates, are nonpathogenic and do not harm the host. Many parasitic organisms may be pathogenic in one individual and nonpathogenic in another. For example, most individuals, infected with Entamoeba histolytica in the digestive tract do not have amoebic dysentery, and, in fact, may have no disease symptoms at all, but in other individuals, the same organism may produce severe disease and even death. This is a matter of individual ecology and immunology which is not well understood.

Theere is no sharp line of distinction between obligatory commensalism and parasitism. Theoretically, Entamoeba histolytica feeding on tissues of the host is a parasite, whereas the flagellate Trichomores hominis, feeding on digested foods before incorporation into host tissue is not a parasite, nor is Entamoeba coli feeding on undigested particles of food and intestinal bacterias. Neither is there a sharp line of distinction between a parasite and a predator. Houseflies and vampire bats bite and suck blood form livestock, and they might be considered either parasites or predators.

Normally, we think of predators as killing their prey in a short time. Man's relation to the cow is predatory when he kills it for beef, but it is parasitic when he milks it for dairy products. There are some relationships, however, in which the distinction is arbitrary, as for example, in the lamprey feeding on a host fish. The lamprey may kill the fish in a few days or a few weeks, depending upon its size in relation to the fish, and may be considered either a predator or a prasite.

Ecologically, one of the most interesting questions about parasitism is the effect of the parasite population upon the host population. The oldest and best adapted parasites have little or no pathogenic effect upon the host. Pinworm infections in children are relatively harmless, though they may cause a minor irritation around the anus. In West Bengal, studies have shown that over 75 percent of the local population harbors hookworm (Necator sp. and Ancylostoma sp.), but there is no demonstrable pathology or effect on the people.

The infectious burden (i.e., the number of parasites per person) seems to be small and insignificant. Most wild animals have parasites are able to maintian excellent health with these parasites. Zebras in Kenya are intensively infected with many internal and external parasites, but they are usually in remarkably robust health. On the other hand, there are numerous examples where parasitic infections are harmful and debiliting to the host. Certainly, hookworm necator americana in the southern United States, as it occurred 50 years ago, was a debilitating disease, sapping the strength and health of the infected person.

Throughout the world, malaria still remains as a major health problem and it causes a great amount of illness and human misery. One of the most important debilitating disease of modern times is schistosomiasis an infection of blood flukes of the genus Shistosoma. This is a waterborne infection and has spread throughout the tropical world with the expansion of irrigated agriculture. In animal populations, parasites may also weaken infected individuals. In the bighorn sheep of Wyoming and Idaho, lungworm infections are still a major cause of illness and mortality. In dogs and wolves, heartworms reduce the animal's vitality and hunting succss. In meadow vole populations, botfly parasitism causes larger spleen weights in infected mice and reduces their survival when exposed to cold temperatures.

Periodic reductions of field voles in Wales have attributed in some years to heavy infections with tuberculosis among the mice, but in other years similar reductions occurred without the infection. Similarly, reductions in populations of red grouse in Scotland have been attributed in some years to the round worm parasite, Trichastrongylosus pergracilis, but this has not a consistent finding in all year. Occasional heavy mortality of gray squirrels in Baltimore has been associated with coccidiosis, caused by a protozoan parasite of the digestive tract. In 1971 a virus disease of wildfowl decimated 90 percent of the pheasants of England, in much the same way as myxomatosis reduced rabbit populations of England, and Europe in the 1950's. Thus, parasitism as an interspecific population relationship may be balanced or unbalanced in regard to its effect upon the host.

A seriously unbalanced relationship, in which the host dies is, of course, selectively disadvantageous to the parasite as well. Another outstanding feature of parasitism is the interspecific complexity which was evolved in some species. Many parasites have primary and secondary hosts and, in some cases, tertiary hosts for various stages of the life cycle. The malaria organism Plasmodium falciparum, for example has parts of its life cycle, the sexual phase of reproduction, within the mosquito, and part, the sexual phase of reproduction within the human host. Schistosomes infect man or other vertebrate animals as the primary host for sexual reproduction. The Chinese liver fluke, Clonorchis sinensis, infects man as a primary host, aquatic snails as a secondary host, and fresh-water fish as tertiary hosts. Despite the complexity of this life cycle, it is a very successful and widespread parasite, infecting millions of people throughout the Orient.

Competition

The word "competition" denotes striving for the same thing. At the ecological level competition becomes important when two organisms strive for something that is not in adequate supply for both of them. Thus, plants compete for light and nutrients in a forest, and animals compete for food and shelter when the latter are relatively scarce in terms of the density of animals. If the population consists of only a few scattered individuals, competition will not be a factor of ecological importance. In the arctic, for example, plants may be so few and scattered due to the severe climate that no competition for light occurs. The result of competition is that both parties (that is, the competition) are hampered in some manner. At the population level this means that density or rate of population energy flow, will be reduced or held in check by the competitive action. There is another type occurs when two organisms interfere with one another while striving for something even though that something is not in short supply. For example, the organisms might secrete substances that interfere with each other, or they might even eat each other.

Many ecologists prefer not to include such direct mutual inhibition under the heading of competition. Both intraspecific and interspecfic competition can be very important in determining the kinds and numbers of organisms. Intra-specific competition is an important factor in those populations that tend to be selfregulated in the manner described above. An interesting behaviour pattern, which results in intraspecific competition for space and a rather effective control of population size, is known as territoriality; it is characteristic of many species of birds and some other higher animals.

At the beginning of the breeding season the male of a territorial species of bird will "stake out" a definite area of its habitat (that is, the "territory") and defend it against other males of the same species, with the result that no other ale is allowed to enter the area. Much of the loud bird song we hear in the spring is for the purpose of announcing to other males "ownership" of land, and not for the purpose of wooing the female, as is often supposed. A male that is successful in holding his land has a high probability of mating and nesting, while a male that is unable to establish such a territory will not breed. Once the pair is formed, the female also joins in the defense of the territory in many species. The defended territory also serves a positive purpose of insuring the complex business of caring for the nest and young will not be interrupted by the presence of other males and females. Finally, it should be mentioned that the actual defense of the territory does not usually involve much actual fighting or other severe stress. Would be invaders respect the established bird; loud songs or threat displays usually are sufficient to cause the invader to withdraw. However, if one of the pair should be killed it is very often quickly replaced by a bird form a reservoir of individuals not established.

Therefore, since the territorial behaviour pattern helps avoid both overcrowding and undercrowding it can be siad to be regulatory in the sense that such behaviour promotes a sigmoid growth pattern, or at least helps prevent an overshoot of carrying capacity. Plant populations, as well as animal populations, may regulate themselves, so to speak, and avoid overcrowding. As one drives through the desert in the southwestern United States one is impressed with the fact that desert shrubs are widely spaced, often almost uniformly distributed as if planted at regularly spaced intervals.

The pattern would seem to be explained by competition for scarce water, which eliminates all but one individual in a given area. However, evolution of self-regulation of the population has occurred in some species, in that severe competition for water is avoided by the production of leaf or root hormones that inhibit development of other individuals in the neighborhood.

Chemical substances rleased by decaying leaves that fall to the ground under the shrub, or by the living root in the soil, inhibit or kill any seedlings that may start to sprout. Such a control mechanism would tend to keep plants spaced apart, thereby reducing actual competition for water, which might result in stunting or death of all plants should there be many trying to grow in the same place. In this kind of "birth control" the quality of the

individual in a limited environment is maximized, not the quantity; is there as lesson here for man?

Amensalism and Antibiosis

Amensalism is an interspecific relationship in which one population is inhibited while the other is unaffected. A simple example is the shading out of certain plants under tall trees. The trees reduce the available sunshine at the ground level, and numerous species of plants cannot find adequate light in the shade. Hence, only shade-tolerant plants with lower light requirements can survive as ground cover in the forest. Antibiosis is a specific type of amensalism. in which one organism produces a metabolite that is toxic to other organisms.

The best known example is the mold, Penicillium, which produces an antibiotic substances causing the death of many bacteria. This was observed and identified in the classic work of Florey and Fleming in England in the late 1920's and early 1980's which led to the development of antibiotics in clinical medicine. Penicillin, strepotomycin, aureomycin and other antibiotics used against disease organisms all represent a similar ecologic principle, and one which is widespread in nature.

Many species of fungi and lichens produce metabolic substances which inhibit bacterial growth. The green algae Chlorella produces substances which inhibit the growth of diatoms of the genus Nitzschia, and conversely, Nitzschia produces a substance which inhibits Chlorella. Many plants also produce substances which are toxic or inhibitory to animals. Old cells of Chlorella produce a substance which inhibits the feeding of Daphnia. Algal blooms of some blue-green and red algae produce chemicals toxic to fish. The well known Red tide of southern coastal waters, produced by the flagellate Gymnodinium brevis, can result in massive fish kills.

Land plants also produce inhibitory substances. The roots of some species of trees produce substances which inhibits the growth of other trees, both of their own species and different species. This may result in a characteristic spacing pattern of trees within a forest, dependent not so much upon competition for sunlight at the conopy level, but upon competition for space, water or nutrients at the subterranean root level. Many plants produce substances toxic to animals which touch them or eat them. In Nevada and Utah the desert bush Halogeton glameratus produces oxalic acid which is poisonous to sheep.

Poisonous toadstools to and tubers have been well known throughout the history of man. In South America and Africa, many native people harvest the tubers of the poisonous manioc plant, but they place it in a series of soaking vats to leach out the toxic chemicals before consuming the starchy tissue of the roots. A small African monkey, the talapoin monkey (Cercopithecus talapoin), has learned to steal the manioc from the soaking vats at just the proper state. Many of these substances are in a class of compounds known as ectorines of environmental harmones. Ectorines are substances produced by the organism which affect other organisms of the same or different species. Not all ectocrines are toxic or inhibitory, of course; many are stimulatory or beneficial. But collectively, they represent chemical messengers by which individuals and species are interrelated.

Aggregations

All organisms react on their habitat in one way or another, and when they occur in numbers these reactions produce a conspicuous effect. Water conditioning occurs hence physical or chemical changes occur as the result of organisms living in it. Compared with unconditioned water, these changes may have either a harmful or a beneficial effect on organisms introduced into the water after the original organisms have been removed. Water is said to be homotypically conditioned when the changes were previously produced by individuals of the same species as being studied and heterotypically conditioned when the changes were produced by a different species.

Experimental studies have demonstrated that gold-fish grow faster in water that has been homotypicaly conditioned for 24 hours than in unconditioned water. Both fish and amphibian larvae also do better in water conditioned by the presence of molluskncs than in unconditioned water. The marine flatworm Procerodes wheatland will survive much longer when transferred to fresh water conditioned by the presence of either live or dead individuals of the same species or by freshwater species of flatworms than they do in unconditioned fresh water.

The longer survival in toxic solutions, faster growth, and greater reproduction of protozoans, snails, flatworms, cladcerans, amphibian larvae, and fish occurring in aggregations rather than as isolated individuals is attributable to water conditioning. Benefits derived from aggregating are shown in other ways by terrestrial animals. In honeybees, when hive temperatures drop below 14°C (57°F) during the winter, they form clusters and maintain a mass temperature several degrees above outside temperatures. This is brought about by increased metabolic oxidation of honey in their bodies and by increased muscular activity.

The compact cluster presents a surface area for heat loss that is less than the total surface area of the individuals separately. When there is danger of overheating, the bees in the hive spread out on the combs and fan with their wings to create a circulation of air. They will also carry water into the hive and place small quantities both outside and inside the comb cells. The forced air circulation evaporates the water and cools the hive. Bees also cool themselves by constantly moving their tongues in and out of their mouths, exposing to evaporation the moisture that is present on them as a thin film temperature regulation is less well developed in other social Hymenoptera.

Allelochemistry

Included here are the coactions whereby chemicals secreted by one organism affects the growth, health, or behaviour of other organisms. Allelopathy is produced in plants when toxins are liberated that inhibit seedling growth in the vicinity. This may affect succession of plant species, especially important in the early stages. Some pioneer species in the abandoned field sere produce substances inhibitory to nitorgen-fixing and nitrifying bacteria. This retards invasion of other species that require higher nitrogen concentration in the soil. Volatile inhibitors are generally more prevalent than water-soluble ones, and relatively more prevalent in arid than humid climates.

Antibiotics produced by bacteria, fungi, actinomycetes, and lichens are widespread in nature and may be one of the reasons why bacteria pathogenic to man cannot mutiply well in soils. A number of antibiotics, such as penicillin, have been used extensively in human medicine. The use of allelochemic coactions agriculture has possibilities but has yet to be exploited.

Among animals, overcrowding of tadpoles in culture dishes in associated with the occurrence of peculiar round vacuolated cells in the intestinal tract and feces that appear responsible for curtailment of further growth. Under laboratory conditions, killer stocks of Paramecium aurelia produce a toxin, paramecin, at the rate of one unit-particle per animals per 5 hours. One unit-particle is enough to kill one individual of sensitive stock of the same species as well as being lethal to other species of Paramecium. Conditioning that becomes unfavourable\ homotypically may sometimes be favorable, or at least tolerable, heterotypically. Thus in protozoan infusions there is a microsere of one species succeeding another.

Allechemic effects are of great variety in both plants and animals: repellants, escape substances, suppressants, venoms, inductants, attractants, signals, stimulants, autotoxins, antoinhibitors, and so on. Pheromones chemical messages between members of a species especially important in reproductive behaviour, social regulation and recognition, alarm and defense, territory and trail marking, food location, as so on Many of these effects are beneficial to the individual; others serve for competitive purposes.

Protocooperation

In the positive relationship between populations, protocooperations is a short step ahead commensalism, to cooperation, both populations gain by the association or interaction of some kind. A considerable work on protocooperation was one by Alee. In nature, crabs and coelenterates are often found associated with each other. This association is of mutual benefit to both. The coelenterates grow on the back of the crabs, or sometimes planted there by the crabs.

The coelenterates provide protection to the crabs. In turn, these are transported by crabs to new situations, which ensure them good supply of food. There is no physiological relationship betweeen the coelenterate and the crab, and none is absolutely dependent upon the other. There exists complex relationships between different populations. Some of these relashionships play a vital role in the regulation of populations. Successful parasitism represents something of compromise between two populations. Predator-prey system is the product of long evolutionary process. A close relationship exists prey, but it is not always the case. The inter-specific competition is a basis to the theory of natural selection and evolution of species. While commensalism, proto co-operation and mutualism beautifully illustrate how interspecific relations can be beneficial.

3 Population Growth

Population studies are fundamental to many human problems in agriculture, pest control, wildlife management, and economic planning. The purpose of population research is to understand the interplay of factors influencing population range, and ultimately to predict and regulate the course of populations. Man obviously wants to limit or reduce populations of organisms, whereas he has wish to maximize within reasonable limits valuable and productive animals for aesthetic or economic reasons. The greatest challenge of all, as well as scientifically and philosophically the most difficult are of human endeavor is understanding and regulation human population change.

The world population was almost stationary during Stone Age, an estimated 10 million people. Sometime between 8000 B.C. and 6000B.C. man learnt the art of growing food, and so could support a large population. As a result the human population increased 50 times in 7000 to 9000 years. In the next two centuries the population doubled and reached an amazingly high figure of one billion. It took just 100 years to double the population to reach four billion in 1975.

Alarmingly, in 12 years it reached the five-billion mark on July 11, 1987. At the present rate of growth we will add another billion in the next 12 years, unless drastic measures are taken. The present world population of 5.43 billion, which is growing at the rate of 97 million people per year will touch 8.5 billion by the year 2025. About 95% of the population growth will be in the developing countries. At present we add about 500 million people in about six years times, a figure which took thousand of years to reach before 1950 AD.

About 3,42,000 babies are born every day in the world and about 1,5,000 persons die, leaving a net increase of 2,07,000. Thus within a week's time there will be nearly 1.5 million more people. The UN population Fund in a recent report says that in the next 10years there will be one billion more people on Earth, making a total of 6.4 billion. That means another China to house, a million more consumers to feed, house, clothes, and educate.

Trends in Population Growth

There is no doubt in the fact that human population are drastically out of balance. For thousands of years prior to the seventeenth century, the population of the world was less than 500 million people. Human population were limited by high mortality and even in the great civilizations of Egypt, Greece and Rome, expectation of life probably did not exceed 30 years.

The class checks on human population were disease, famine and war. In the fourteenth century, at least 25 percent of the adult population of Europe died in epidemics of bubonic plague and between the years 1348 and 1379 A.D. England's population was reduced almost 50 percent, Cholera, typhus, small pox, malaria yellow fever and sleeping sickness have all been among the great killers of man throughout history. Several of these diseases, especially malaria and yellow fever, severely limited the abundance and distribution of people in tropical countries until the advent of modern medicine and public health. Asia has traditionally been the great stage for famine though starvation; death also occurred in Europe. The famous Irish famine of 1850 resulted in the death of more than 12 percent of the adult population between 1846 and 1851. In China, in 2000 years preceding the 20th Century, over 1800 separate famines were recorded, many of them killing million of individuals. More recently, 4 million famine deaths occurred in China in 1920-21 more than 5 million is Russia from 1918 to 1923; and over 2 million in India in 1943.

Almost agriculture and transportation have alleviated famine mortality, malnutrition continue to be significant demographic for, particularly since it accentuates illness and mortality from infectious disease. War has also been a major mortality factor throughout history,not only by causing direct battle casualties, but also by stimulating disease and famine. Great epidemics and massive crop failures, have often been triggered by the disruptive condition of warfare. The casualties of war have frequently involved as many civilians as soldiers. Some historians believe that the Thirty year war in Europe between 1618 and 1648 resulted in the death of 30 percent of the inhabitants of Germany and Bohemia.

Despite the combined action of disease, famine and warfare, several major development in the sixteenth, seventeenth and eighteenth centuries stimulated population growth and eased the heavy mortality burdens of man. The opening of the New World provided a vast new area of space and nature resources of economic exploitation. The industrial revolution greatly increased the productivity and control of infectious diseases.

The industrial revolution greatly increased the productivity and mobility of man. These factors combined to accelerate human population growth and the world population curve began to rise upward.

The following table indicates the populations growth rates

Region	Population (billions)			% Annual Growth rate		
	1985	2000	2025	1950-85	1985-2000	2000-2025
WORLD	4.8	6.1	8.2	1.9	1.6	1.2
Africa	0.56	0.87	1.62	2.6	3.1	2.5
Asia	2.85	3.55	4.54	2.1	1.6	1.0
Latin America	0.41	0.55	0.78	2.6	2.8	1.4
North America	0.26	0.30	0.35	1.3	0.8	0.6
Europe	0.49	0.51	0.52	0.7	0.3	0.1
U.S.S.R.	0.28	0.31	0.37	1.3	0.8	0.6
Oceania	0.02	0.03	0.04	1.9	1.4	0.9

(Source: Our common Future, 1987.)

Between the birth of Christ and the seventeenth century, the world's population took approximately 1600 years to double.

The next doubling as achieved within 200 years, and the next from 1 to 6 million people within 80 years, from 20 to 4 million people were completed by 1975 a span of 45 years and at the current rates of growth the world population will double again within another 35 years. Currently the world's population is increasing at approximately 2 percent per year—a net increase of about 1,300,000 per week. Every day of every week, the world's population increases by more the 207,000 people. All this is primarily a product of man's increased life span a reduced mortality through modern medicine and agriculture. Man's fertility patterns were basically established during periods of much higher mortality rates thousands of years ago and they have remained high in the intervening year.

Global Distribution

The developing countries least able to afford rapid population growths are experiencing the greatest growth. The population is not evenly distributed around the world. More than half the world's population regions of the total world, i.e. Europe, North America, former USSR and Oceania had only about 22% of the total world population in mid 1988. The remainder was distributed in the developing region of the world. The growth rate is most rapid in Africa. In the coming decade, the population there will rise from 640 million to 950 million and it will be the highest regional growth rates the world has ever known.

According to projection, countries like Ethiopia, Nigeria, Zaora and Kenya will double their populations within the next 20 years. In North Africa the growth rates are slower, but th figures are still rising fast. South Asia and Africa together will account for over half of the increase. Southern Asia's population will rise from 1.2 to 1.6 million. In Eastern Asia the population growth rates have declined rapidly in the past decade, but the number will still increase from 1.3 to 1.5 million people.

Several Arab countries have very high population growth rates and the 12 countries in West Asia will increase in their population to 131 million by 2001. The population; of Latin America and the Caribean is projected to increase by 100 million from 440 to 540 million. In Central and South America, most countries have annual growth rates in excess of 2%.

Dependency Ratio

In addition to the large concentration of population in developing countries, the percentage of population under 15 years and over 64 years of age what is called the dependency ratio is also very high. The population under

15 year is about 45% in Africa, 35% in Asia and 39% in latin America, whereas it is only about 23% in North America 22% in Europe and 29% in Oceania. However, the percentage of population over 64 year is higher in the developed regions of the world. North America has about 11% of its population over 64 years.

Europe has about 13% and Oceania about 8%. It is only 3% in Africa and 4% in Asia and latin America. The high dependency ratio in the developing countries means that age group between 15-64 has to support a large population with requirements for housing, clothing, education, health and other social service. The large young population in developing countries also has demographic implications in the near future.

Asia and Africa, and the regions which will account for about 75% of the world's population in the next 35 years or so, and will have the greatest proportion of poor people. In South Asia 51% of the population lives below the poverty line and in sub Saharan Africa 45% Malnutrition, ill-health, illiteracy and the lowest life expectancy, accompanied by inadequate technology and economic poverty, are predominant in South Asia and Africa. It is not very difficult to imagine the miserable plight of the people in their regions in the next century. The UNEP has warned that continuing high population growth will result in continuing poverty and increasing environment damage and will lead to higher levels of urban growth and international migration.

The population report support this that it will be impossible to keep pace with population trends. If the population growth continues as expected, it says within 10 years there will be more than 200 million children who will receive no education. What does this mean in terms of human welfare and environment quality? It usually means that many countries have excessive demands for products and services which cannot be met.

It is exceedingly difficult, if not impossible, for those countries with the highest rates of population growth to meet the requirements of their people for food, housing, jobs and the physical amenities of life, and it is virtually impossible for them to maintain adequate services in education, medicine, public health and social welfare.

Thus we emerge with such problems: (1) 60 percent of the world's people are still inadequately fed, receiving a percapita calorie, intake less than that recommended by the UN (2) 80 percent of the children of India shows evidence of malnutrition; (3) there are only 9 acres of land surface available for every person on the earth, and only 26 percent of this is available for agricultural production; thus only 1.2 acres of potential cropland are currently available for each person and this is decreasing each year (UN States of Hunger Report); (4) many countries, especially in Asia and Africa, still have an illiteracy rate exceeding 70 percent; (5) in many developing nations, there is only one physician for every 5,00 to 10,00 people, a ratio which makes adequate medical care impossible.

This list could be extended considerably to indicate that many parts of the world are currently living under severe population pressure. While the western nation gain strikingly in prosperity, most of the world's people face increasing poverty and population pressure. When to global picture is considered, the outstanding fact exists that we can not provide a quality life or maintain a quality environment for the majority of the world's people at the present time, let alone provide for more than 1,600,000 new people every week.

Ecologically, this represents a highly unstable condition, it cannot persist indefinitely without major corrective or compensating changes. These grim facts coupled with increasing global communication, rapid social and technological change and rising expectations throughout all nations add fuel to the fire of social and political instability. As noted by Eric Hoffer in The Ordeal of Change:

> "...... a population subject to drastic change is a population of misfits-unbalanced, explosive, and hungry for action. In other words, drastic change under certain conditions, creates a proclivity of fanatical attitudes, united action, and spectacular manifestation of fluting and defiance. It creates on atmosphere of revolution.

The next decade may decide the future of the earth as hesitation for humans. Already our impact has been sufficient to degrade the soil of millions of hectares to thin the ozone layer threatening the rain-forests and to initiate global warming the full consequences of which cannot yet be calculated. At some point in the not too distant future the changes in the environment may cross the threshold of catastrophe. Again we see the intimate relationships between ecology and human affairs. It does not imply that ecology can provide the answers to all these problems but surely a failure to recognize their ecology background invites certain disaster.

Indian's Population Situation

India's population, as on March 1, 1992 stood at 843.93 million(437.60 males and 406.33 million females). The second largest population country, India is now the home of 16 percent of world's population. With this, every sixth person in the world is now an Indian. The country however accounts for 2.42 percent of the total world area. Population counts in India in the modern sense started in 1881 since then, a population census has been conducted regularly every ten years.

A census of population was conducted during 1871-72 but it was nosynchrononus. Moreover, the 1972 census omitted served territories, the population of which totaled some 33 million in 1881. The population of India as recorded at each decade census form 1901 has grown steadily except for a decreased during the decade 1911-21. In absolute terms the country population has increased by 160.60 million from 680.3 million in 1981 to 843.9 million in 1991. This is ten times the population of Australia and is more than the population of Japan.

An encouraging feature revealed by the census result is the decline in the growth rate of the population in the last decade compared to 1971 to 1981. It marginally decreased from 24.66 percent recorded in 1971-81 to 23.50 per cent during 1981-91. In most state the growth rate declined during the decade. However, Andhra Pradesh, Arunachal Pradesh, Madhya Pradesh, Maharashtra, Nagaland, Jaipur, West Bengal, Daman and Diu, Lakshadweep and Pondicherry which account for one third of the country's population recorded increase in growth rate. Nagaland registered the highest growth rate of the 56.86 percent while Kerala registered the lowest 13.98 percent.

Uttar Pradesh continues to be largest state population wise with 16.44 percent people of the country living there (138,760,417), followed by Bihar with 10.32 percent (86,338,853) and Maharashtra with 9.33 percent (78,706,719). West Bengal occupied fourth place accounting for 80.06 percent (67,982,732) of the country's population followed by Andhra Pradesh with 7.86 percent (66,304,854) Bombay continued to be most populated city of the country with urban population of 12.57 million, Calcutta taking the record place having 10.86 million, Delhi ranks third with a population (urban) of 8.38 million followed by Madras 5.36 million.

Population Density

One of the important indices of Population concentration

is the density of population. It is defined as the number of person per square kilometer. The population density has gone up from 216 in 1891, to 267 person in 1991. In 1901 it was 77 person. Density is the highest in Delhi (6.319) followed by Chandigarh (5,620). Lakshadweep which has the smallest population size is the third most densely populated area with a density of 1,615 person per square kilometer , Pondicherry ranks fourth (1605) followed by Daman and Diu (906) Arunachal is the most sparsely populated area with a density of 1,615 person per sq kilometer.

The ten heavily populated districts of the country are Calcutta Madras, Greater Bombay, Hyderabad, Delhi, Chandigarh, Maoe, Haora, Kanpur City and Banglore. All of them have density of above 2,00 person per sq km and 5.0 percent of the country's population lives in these districts. The average density of these 10 districts. The average density of these 10 districts is 5,791.

Sex Ratio

In any study of the population, analysis of the sex composition plays a vital role. The sex composition of the population is affected by the differentials in the mentality conditions of males and females, sex relative migration and the sex ratio at birth.

Table:

Sex Ratio (Females per 1,000 Males) in India, 1901-1991

Census year	*Sex ratio*
1901	972
1911	964
1921	955
1931	950
1941	945
1951	946
1961	941

1971	930
1981	934
1991	929

The sex ratio is defined as the number of females per 1000 males. In India it has generally been adverse to women. The ratio has also declined over the year except in 1981 when it slightly improved to 934 from 930. In 1991, there has been a fall by five points to 929 per 1000 males. But Kerala represented a different spectrum. The State has a higher number of females than males, 1040 females for 1000 males.

In Indian context, a sex ratio of 950 or above can be considered as favorable to females. The states and Union Territories coming under this category besides Kerala are Himachal Pradesh (96), Andhra Pradesh (972) Goa (969) Karnataka (960) Manipur (961) Orissa (972) Tamil Nadu (972) Dadar & Nagar Haveli(953), Daman & Diu (972) and Pondicherry (982) Chandigarh accounted for lowest number of females per 1000 with 793.

Demographic Transition in India

Many scholars have inferred from the actual experience of several European countries and the United States of America that populations historically pass through different more or less well-defined stages in terms of their levels of fertility and mortality. As a generalized explanation of process of mortality and fertility decline in those countries, the transition theory refers to a stage of high fertility in those countries, the transition theory refers to a stage of high fertility and high mortality in simple, mostly agarian, societies which finally transforms itself to a condition of low fertility and low mortality after passing through several other stages.

Today the transition theory it sought to be equally applicable to the less developed counties which are still in the early stages of demographic change. In effect this theory has come to be increasingly regarded as theory that might anticipate the future demographic trends in countries which are currently having high population growth due to rapidly decreasing death rate and high but more or less stable or only slowly declining birth rate.

There are several expositions of demographic transition theory, but the one dividing the total period into five stage is more commonly used. These stages are as below;

(a) The high stationary stage characterised by high birth rates and high death rates;

(b) The early expanding stage with high birth rates but declining death rates;

(c) The late expanding stage with falling death rates but more rapidly falling birth rates;

(d) The declining stage with low mortality now natality and deaths exceeding births.

The first stage which is characterised by agrarian low income economy is marked by relatively stable birth rate at the maximum fertility level found in societies not using contracepitives, and the death rate is found to fluctuate in response to conditions of prosperity or famines, epidemics and other disasters.

As the economy progresses to become more interdependent specialised and market dominated with the generation of agricultural surplus, the average death rate begins to decline under the impact of better organisation and improved medical knowledge and care; however, the birth rate continues to remain almost stable at the earlier level. This leads to an increasing growth rate of population which is the early expanding stage. As the death rate continues to decline with further improvements in living conditions birth rate also starts declining first slowly and somewhat later at a faster speed making the 'late expanding stage'. In this stage the growth rate starts getting smaller and smaller than what it was in the early expanding stage.

In the fourth stage when further reduction in death rate becomes harder to attain, the death rate stabilizes at a low level(between 8-12 per 1000 persons), the birth rate as approaches equality with the death rate and a more gradual growth rate is established when both birth and death rates are low but birth rate fluctuates some what from year to year in response to voluntary decisions of the people. In the last stage of demographic transition which started to appear in recent years in a few European countries like Germany, Sweden, Switzerland etc., the birth rates has declined even below the replacement level and there may not be enough people in the next generation to replace the present one. For India, we now have a history of around 100 years of changes in birth and death rates.

Table

Birth and Death Rates in India from1891-1901 to 1989

Year	Birth rate	Death rate
1891-1901	48	48
1901-1911	49	49
1911-1921	49	49
1911-1931	47	37
1931-1941	45	33
1941-1951	43	31
1951-1961	44	26
1961-1971	42	20
1971-1981	37	15
1980	33.7	12.6

1981	33.9	12.5
1982	33.8	11.9
1983	33.7	11.9
1984	33.9	12.6
1985	32.9	11.8
1986	32.6	11.1
1987	32.2	10.9
1988	31.5	11.0
1989	30.5	10.2

The first stage of demographic transition in India continued till about 1920 when both death rates and birth rates were high. This stage was also marked by purely agrarian economy with hardly any significant industrialisation.

High mortality levels were hardly and for the variable growth rates of the population during the 30 or more years before 1921. A severe famine affected large areas of the country in 1896 and 1897. In the Bombay Presidency (excluding the princely states) especially, the effects of the famine were aggravated by a severe plague. As a result of the two scourges, the 1901 census recorded a population loss of about 2 percent from the 1891 level of 18.8 million.

Like the 1891-1901 decade the next decade also witnessed several local famines and a severe one in 1907 in most parts of Uttar Pradesh. Plague was in evidence in the Bengal and Bombay Presidencies, and both plague and malaria were widespread in the Punjab and Uttar Pradesh where population growth was negligible. Yet, because the country as a whole suffered from famines less widely and for shorter durations, there was an appreciable rise in the population growth rate compared to the previous decade. During the 1911-20 decade India suffered from an influenza epidemic that caused an estimated 7 percent of the total population to die and led to an absolute decline in the population of the Indian Republic (according to the present boundaries) by 0.77 million at the 1921 census.

The second phase of demographic transition in the country started with the beginning of the 1920s and has spread well up to 1971. This is also the period when industrialization began in the country and took some roots. During this period the major causes of high mortality-famines and epidemics-have been gradually brought under control and between 1921 and 1951, the country witnessed a gradual rise in population growth rates. The decline in death rate became sharper during the 1950s without any appreciable decline in birth rate resulting in a population growth rate of 1.96 percent per annum during 1951-61 in comparison to 1.25 percent during 1941-51 decade.

During1961-71 the birth rate still remained quite high and due to further decline in death rate the growth rate increased to 2.05 percent. India seems to have entered the third stage of demographic transition since 1971 when birth rates has also started declining. During the birth rate leading to a plateau in the population growth rate during the 1960s and the 1970s. In the 1980s, probably the decline in birth rates was slightly greater than that in mortality; consequently the growth rate during this period is slightly lower than in the previous decade. It is now hoped that there would be faster declines in birth rate than in death rate in the next couple of decades. It is also hoped that India will be at the point by about 2020 if not earlier.

Urbanization

Only one in ten people lived in cities when the century

began more than half will by the century's end. Today the vast bulk of urban population growth occurs in developing countries. The populations of Third World cities, now doubling every 10-15 years, overwhelm government's attempts to provide clean water, sewage, adequate transport and other basic services.

Cities in Industrial countries are gaining people much more slowly, yet even as their growth rates slow or core population decline, these urban areas continue to spread out ward. Industrial and developing countries alike, chaotic, uncontrolled urban growth-whether measured is risking numbers or in the amount of space humans spread out upon—draws on ever more land, water and energy from surrounding regions to meet people's needs. The way cities physically evolve—and the way their development is planned has profound impact on human either recognize the limits of the natural environment or it can destroy the resources on which current and future societies depend, it can meet people's needs equitably or it can enrich some while impoverishing or endangering others.

The world needs an urban planning ethic that is sensitive to these environmental and human dimensions. According to 1981 census, India's urban population was 160 million ranking fourth among the countries in the world in terms of absolute size of the urban population, the first three being China, U.S.S.R. and U.S.A. Though the level of urbanization measured as the percentage of urban to total population has increased from 11 percent in 1901 to only 23 percent in 1981 the number of people living in urban areas and multiplied six folds during the period from 26 million, constituting more than one third of the country's population of 926 million at that time.

The growing chaos in urban areas are due to rapid immigration rather than natural increase. A total of 15.7

million persons have migrated from rural to urban areas in the country during the decade 1971-81. The immigrants in the reverse direction i.e. urban to rural numbered. 6.4 million giving rise to a net addition of 9.3 million people in urban areas.

There are 218 class I cities in India with more than a lakh population each and these account for about 60 per cent of urban population. Among these, there are 12 metropolitan cities viz., Calcutta (9.16m), Greater Bombay (8.22m), Delhi 5.7m), Madras (4.27), Bangalore (2.9m), Kanpur (1.68m), Pune (1.68m), Nagpur (1.29m), Lucknow (1.0m) and Jaipur (1.0m). With the data on class I cities and the emergence of new towns on the eve of the 1991 census, we can examine the likely level of urbanization in India. It is well known that the share of Class I urban population has constantly agglomerations and cities in the total urban population has constantly increased over time. If one looks at the total urban population has constantly increased over time. If one looks at the trend in India (excluding Assam) from 1951 to 1981, this share increased from 44.7 per cent to 60.8 per cent, the increase being 6.8 per cent during the 1970s.

The increase in the share was higher during the 1950s since a very large a number of small towns (857 in all) were declassified on the eve of the 1961 census due to introduction of a restrictive definition of urban areas. In contrast, there was a net addition of 895 towns (or 782 urban agglomerations and towns) between 1971 and 1981 which involved reclassification of almost 250 towns declassified on the eve of the 1961 census. Addition of a large number of small and medium towns during the 1980s brought down the pace of increase in the share of population of Class I cities in 1981. As the number of new towns indentified on the eve of the 1991 census is quite large, one may expect a substantial increase in the population of medium and small towns in 1991. If, however, a large number of them appear as part of urban agglomerations of Class I cities, their population will be counted as population of Class I cities. An examination of the list of new towns and of cities and urban agglomerations indicate that a large number of new urban settlements have appeared as part of urban agglomerations.

Accordingly, we may expect the share of the population at 65 per cent. This is in view of the fact that 80 Class I cities have been added during the 1980s. We may also not forget that the decadal growth rate of Class I cities slightly declined over the decade compared to the previous decade. Assuming that the proportion of population of Class I cities in the total urban population of 1991 is 65 per cent, we get the total urban population as 213 million excluding that the Assam and Jammu and Kashmir. If we add the 1991 population of Class I cities of Assam, the estimate of total urban population would remain at the 1981 level (that is, 60.5 per cent), the 1991 urban population (excluding Jammu & Kashmir) works out to 230 million, almost the same as projected by the Expert Committee on Population projections in 1984. Thus, the urbanization rate in 1991 varies between 25.6 per cent to 27.5 per cent.

Keeping in view that the number of metropolises has increased from 12 in 1981 to 23 in 1991, and that the number of Class I cities has increased from 216 to 296 during the same period, one may expect a higher share of Class I cities in the total urban population than 65 per cent. If this is found to hold good when the data on rural and urban population are released, the total urban population would be less than 214 million. Population in the twentieth century has exploded particularly in the developing countries.

Infact, before 1800, no country was predominantly urban. But by the year 2000, over 50 per cent of the world's population will probably live in urban places and there will probably be more than 250 cities of over one million inhabitants. There has been a world wide trend towards urbanization, the process characterized by the movement of people from rural to urban settlements, from small towns to large cities, and from large cities to the suburbs. This immigration is caused by and in turn leads to profound societies change. Rural areas are depopulated, while urban areas towns, cities and urban zones—become more densely populated. Since most of the cities are growing in an unplanned way there arises a number of problems such as housing, traffic, shortage of open space & pollution.

Problem of Housing

The Indian cities are growing both vertically (multi-storey building culture) and horizontally (invasion of the peripheries). In spite of cities expansion there is growing shortage of house (3.8m), growth of slums and squatter settlements, creating interpersonal and intercity imbalances, breaking down services and acceutuation of the problem of urban renewal. Slums are now becoming permanent features and are proliferating and sheltering larger population every year. They are having small tenements, narrow lanes, lack of sanitation, water supply, electricity, unhygienic environment, etc.

Due to realization of the futility of spending money on their clearance, the Government had decided to accord them recognition and gave their residents whatever amenities the funds allowed. Bombay is estimated to have the largest slum population of 3.3 m followed by Calcutta 3.2m, Delhi 2.6m, and Madras 1.8m. With an unabated inflow of job seekers, these cities are expected to see continuing rise in their slum population. By 1990 when Calcutta will have a population of 12.5m and its slum will be harboring 4.4m Madras population will be

11.8m, 9.7m and 6.0m and of them 4.1, 3.1 and 2.1m slum dwellers. Amongst States, Uttar Pradesh will have the largest slum population of 6.5m followed by Maharashtra 6.2m, West Bengal 4.9m, Tamil Nadu 4.2m, by the end of the decade. The Government's initial strategy was to clear the slums and rehabilitate their dwellers. This had to be given up by the end of the sixth plan as there was constant shortage of funds and developed land.

The emphasis was shifted to an on going central scheme known as "Environmental Improvement of Urban Slums," envisaging a package of services for the immediate removal of unhygienic environment of the slums. The services included drinking water, community bathrooms, street-lighting, sewerage, storm water drains and street paving.

Traffic Problems

The traffic problem in urban areas is one of the greatest problems of the day. Most of the cities and towns are growing in unplanned way. Due to lack of proper enforcement agencies encroachment in all parts of the cities is going rapidly and due to political interference, authorities become in competement. Most of the roads are narrow, haphazard and ill-maintained.

Traffic problem is getting bad to worse because of high density of population desire to have personal means of transport and to live near the city, mixed traffic, lack of civic sense, ill-maintained road, inadequate expansion in the existing road network poor regulation of traffic, etc. Delhi had 12,000 vehicles in 1951 and the number had shot upto 9 lakh in 1986 and this is expected to reach 40 lakh by 2001. Apart from the fast moving vehicular traffic, the problem on the city roads is being further aggravated because of 50,000 or more slow moving vehicles such as tongas, rickshaws and bullock carts and 20 lakh cycles. There is also a growing trend of road accidents. There were 6,037 reported road accidents of which 1,162 persons lost their lives in 1986. According to National Transportation Planning and Research Centre, Trivandrum in every four minutes there is one road accident in India, killing or injuring one person on an average.

Shortage of Open Space

Due to increasing commercial value of the land in cities, open spaces are diminishing, even though green spaces are the lungs of the cities, the guarantors of a healthy and aesthetic environment. In Japan, urban parks are classified under two main categories; parks with structures or city parks and zoned green spaces, which are further subdivided into green space conservation zones, agricultural green space and special historical land space zones/scenic beauty conservation zones.

City park land includes playgrounds, open spaces, planting strips along side road and other lands awaiting full development. It is commonly said that these parks become the pursuit of undesirable activities including sex, crime, gambling, etc. In spite of these negative aspects, it provides thousand times much needed benefits for the city dwellers.

There is a concept to have at least 20 per cent area of the cities under open space in the form of green belts, parks, playgrounds, etc. But it has become an utopian concept. In 1973, master plan of Allahabad was approved and much of the areas were put under green belts. But due to illegal spraw of colonies most of that very areas were converted into concrete belts. The green belts concepts are also not getting proper place in the city environment improvement scheme because of lack of financial support, public interest and professionally qualified staff. Besides this, there is not definite framework for parks and recreational areas in relation to national goals. City parks and green belts are developed mostly by Municipalities or Corporations, Social Forestry and Horticulture departments. Private organizations, voluntary organizations, industries should also come forward to set city parks and green belts. Vacant lands acquired by the Corporation are to be sued not only for commercial or residential settlement but also for establishing big parks.

Local communities must come forward to participate in the management of parks and can also raise funds for it. Country parks can be established in rural areas to provide a better place for recreation and enjoyment of rural people. In addition to parks and gardens, landscaping measures should be applied on flyovers, overhead bridges, wall and other structures, such as the growing of creepers, climbers, scrubs and transplanted trees. There is a need for making legislation for City and Country Parks Act for the forceful implementation of the concept of green belt.

Urban Planning Around the Globe

Virtually all cities share some land related concerns, such as congestion and pollution from motor vehicles, lack of affordable housing, and the cancerous growth of blighted district. Of course, prescriptions for better urban planning are not the same for all parts of the world.

Many land use issues in industrial countries seem mainly concerned with the quality of urban life, and those in Third World cities are often questions of life and death. But all cities, whether surrounded by affluent suburbs or makeshift shanty towns, now need to plan land use far more carefully than in the past before the developing world's urban crises than into catastrophes, and the industrial world' urban crises turn into catastrophes, and the industrial world's problems become issues of survival. The extent to which cities control the use of land varies widely. Japan and Western Europe have the world's most comprehensive urban land use controls; their cities are generally compact, laid out for efficiency in transport, water and energy use, and building materials. Japan protects farmland with strict zoning and tax policies. Its suburbs are of moderate, not low, density.

Much of Japan's urban activity is concentrated in Tokyo and along the dense Tokadio corridor between that city and Osaka. This compact, linear form has allowed great efficiency in transport, including development of the world's first high-speed train, which has been running since 1964. However, the country's 33 million automobile—nearly twice as many as in all other countries combined—still plague Japanese cities with extreme traffic congestion, especially in the Tokyo area. Western Europe has a long tradition of activity controlling land use so that the small amount of available space serves the public's interest more than that the private developers. Paris, a city for nearly 2,000 years, has planned and regulated land use since the Middle Ages.

Many of England's urban areas, still fulfilling a far sighted decree by Queen Elizabeth I in 1580, are ringed by green belts intended to protect farmland and prevent sprawl. But in post war decades the automobile has had a profound effect on Europe's compact character, fuelling suburbanization. In the fifties and sixties, some older European cities embraced the U.S. model, demolishing corner shops and cafes to make way for urban freeways. Later, planners tried to contain suburbanization by diverting growth into smaller satellite towns. Since the mid-seventies, they have turned their attention to revitalizing shrinking inner cities.

Control of land use in Eastern Europe and the Soviet Union has not been as effective as might be assumed for centrally planned economies. Urban planning therefore is not well coordinated with economic decisions, and sometimes reflects rivalries among government ministries. Extreme water air, and soil contamination from poor placement and regulation of polluting industries is a critical concern.

Urban planners in these countries historically have supported public transport and promoted urban densities high enough to make it viable, in recent years and use has become less efficient. Nearly all new housing has sprung up at the edges of cities, with few accompanying shops, jobs, or schools, resulting in excessively long commutes. On going political reform and the region's cities will become more automobile-oriented, perpetuating the environmentally destructive patterns of western industrial cities. Among industrial regions, North America and Australia have the weakest planning traditions. Governments on these continents have done relatively little to guide development beyond separating industrial areas from those zoned for commerce and housing.

Many have enforced low-density zoning in an effort to curb urban expansion—a move that, ironically, achieves just the opposite. Zoning codes that restrict residential density, usually by requiring each house to occupy its own large lot, have forced development to consume even greater tracts of open space. And to accommodate new, outlying communities, roads have been extended further outward, attracting even more spread out growth.

Even older U.S. cities are becoming more dispersed now than in the pre-automobile area. Growth in the New York metropolitan region has turned to sprawl; while population has grown only 5 per cent in the past 25 years, the developed areas has increased by 61 per cent—consuming nearly quarter of the region's open space, forest, and farm land. Many people move out to the suburbs and exurbs seeking open space and bounds with nature that come only in a rural setting.

Most of these residents continue to maintain urban lifestyle—communiting to jobs in the city and demanding urban amenities from a suburban shopping mall near their home. The result is neither urban nor rural living but a destructive compromise that the environment sustain. Developing countries have the loosest controls over how cities develop. Local governments in Asia, Africa, and Latin America often have neither the authority to guide land use nor the funds to provide basic services. Municipalities are fiscally dependent on central government, which are themselves struggling under crushing debt burdens. With few exceptions, urban planning is a relatively recent phenomenon in developing countries, even in the capitals. In India it reportedly began with Delhi's Master Plan in 1962. In Nigeria, the first systematic effort to plan cities did not begin until 1971; the capital, Lagos, still has no designated metropolitan government and no formally defined urban area. The Lagos State Ministries that have their own priorities.

The Third World is burdened by several enormous, rapidly growing cities—including-Sao Paulo, Shanghai, and Mexico City—whose sheer size and instability create problems on an entirely different scale. As early as 1964, researcher Ronald Wraith used a single word to denote these giant cities racked with pollution and rimmed by shanty-towns: "Megalopolis-the city running riot with no one able to control it." Since then, megacities have become increasingly characteristic of the developing world. In 1950, only three of the world's 10 largest cities were in the Third World by 1980, seven of them were there. Although skilled land use planning is badly needed in these megacities, its effectiveness is limited in the absence of other kinds of change. One reason is that most of the physical growth takes place in illegal unplanned squatter settlements, rendering useless even existing mechanisms for guiding land use. Called fevelas in Brazil bidonvilles in French-speaking West Africa, Spanish in the Middle East, and kampungs in Indonesia, these illegal communities hold 30-60 per cent of the population of many Third World cities.

The unhealthy conditions in these settlements can only be addressed fully through extensive economic and social reforms that attack the root causes of poverty not just in cities but also in the rural areas that urban migrants abandon in search of economic opportunity. The effectiveness of land use planning is also limited in the absence of family planning programs. (Only in Africa does migration account for the increase in much of Asia and Latin America results from the fertility of people already in cities). Finally, planners in the Third World's giant cities faces colossal environmental problems, from deadly air pollution in Mexico City, to indiscriminate dumping of toxic wastes in Alexandria, to the actual sinking of cities such as Bangkok, Jakarta, and Shanghai due to over-drawing of ground water.

Crowding

Crowding is a conspicuous ecologic force impinging upon modern man. It results from more than a simple increase in population density. It is also a product of activity patterns psychological needs, and social contact. Crowding implies a force, a pressure and psychological reaction. No clear definition of crowding is possible at the present time; it remains an operational concept with many subjective qualities.

Crowded population are, however, easier to recognize than to define, and they are studied more carefully, better definitions may become possible. For the time being, we might define crowded populations as those in which there seem to be excessive number of individuals per unit space in relation of the activity of the industrials and the quality of the environment. With an operational concept of this type, ecological studies have shown that some rodent populations may be "crowded" in relation to their social behavior and habitual requirement at a density of 5 individuals per acre, where as other at the same species may not be "crowded" until they have a density of several hundred per acre. Species differ in this regard, and so also individuals and social groups within the same species.

A western cattle rancher may feel "crowded" at a density of 10 homesteads per square mile whereas a modern suburbanite may feel relatively "uncrowded" at a density of several hundred homes per square mile. The western rancher ranges several square miles per day over what he considers his personal property and he does not welcome potential competitors in his territory. Conversely, suburbanite restricts his daily movements, apart from going to work and to shop, to a lawn around his home, and he does not view neighbors as a competitive threat. Hence, in both animal ecology and human sociology, it is important to distinguish between population density per se, and crowding, since the latter depends so much on behavioral factors.

The spatial requirements of man are obviously very variable, and result from both individual difference and social conditioning. It is impossible to state what the optimal spatial and social needs of man are as those of the Japanese city dweller ar much different than those of the Kansas farmer. Nevertheless, it is safe to assert that man is now subjecting himself to increased crowding through the combined forces of population growth, urbanization, transportation and mass communication. Man's perception of crowding has skyrocketed upwards in recent years. The growth of urban traffic and the increased number of people in public buildings, schools, hospitals, stores, exhibitions, recreational areas, etc., have all contributed to the feeling of population pressure. Through the mass media or audios, television and newspapers, individuals are vastly more aware of worldwide events. Social pressures at any point of the globe are quickly felt by all nations. Local crime, violence, and warfare now become worldwide worries in a matter of hours.

The people of one continent feel threatened by those of another continent thousands of miles away. The inevitable result has been a great magnification of social anxiety and population pressure. The world has thus become more limited in its psychological space, and man more crowded in a social and behavioural sense. How does one evaluate crowding as an ecologic force? It definitely requires knowledge of at least three major parameters: (1) population density, (2) environmental structure, and (3) social behaviour in relation to each of the possible in both animals and man, but in both, especially man, it is often possible to isolate crowding as the key variable.

Many other factor are interwoven. Crowded human populations are often so characterized by poverty, malnutrition, lack of educational and recreational opportunities, varies environmental hazards and unstable social patterns that it is virtually impossible to separate crowding itself as the one key variable. The essential question becomes, how does one measure and analyze crowding per se apart from all it related phenomena? It may also be important to distinguish between crowding of various origins. Since man often seeks social contact and stimulation, does not crowding contain constructive and creative social elements, or is it inevitably a factor in increased social problems and environmental deterioration? From man himself the answer to these questions are confusing and uncertain. The behavioral sciences of man have just begun to investigate human spatial behavior.

More direct clues have come from animal studies where experimental work has been possible. The simpler social systems of many animals have permitted more direct observation of behavior, more ready identification of variable, and more penetrating analyses of causes and effect relationships. In animal studies it has been possible to distinguish between extrinsic crowding; that crowding imposed on the animals as an external force, as in captivity, compared to intrinsic crowding in which dense aggregations result from the social behavior of the animals.

A similar distinction is sometimes possible in human populations, but frequently it is difficult to discern if people are crowded by choice or the external pressures of economics and society. It was demonstrated more than 30 years ago by Professor W.C. Alee and his students of the University of Chicago that most species of animal have optimal levels of crowding, above which and below which deleterious effects occur. These studies showed that certain levels of crowding are necessary to maintain normal biologic function and social activity. Goldfish, for example, grow most rapidly and live longest in social groups of certain densities compared to isolated fish to fish in smaller groups. Fresh water shrimp survive longest and tolerate harmful environmental conditions most successfully at certain optimal stage of crowding.

Similarly bobwhite quail survive harsh winters best and reproduce most successfully in the spring in conveys of certain optimal size. Beyond optimal levels of crowding, however, benefits are quickly lost and various forms of pathology, occur. Growth may be stunted, life span shortened and various diseases or abnormal behaviors appear. Most studies on crowded animal populations in the last 20 years have concentrated primarily on the harmful effects of overcrowding. Such effects usually involve both physiologic and behavioral abnormalities. The physiologic effect of overcrowding centre around endocrine imbalances of the adrenal pituitary system. These imbalances can lead to a great variety of disease condition including gastric ulcers, hypertension, nephrosclerosis, arteriosclerosis, and increased susceptibility of infectious diseases.

Prolonged exposure to crowding and its associated stresses have produced additions disease (hypoadrenalism or adrenal exhaustion) myasthenia (muscular weakness), and a variety of serious metabolic disorders leading to shock, coma, and even death. It should be pointed out that these stress effects are quite variable, however, and do not consistently reflected the same patterns in different individuals and various social groups. Behavioral pathologies also appear in experimental animals as a result of crowding and these are sometimes manifested more quickly than are the physiologic changes.

Overcrowded animals often show increased aggression and violence, abnormal sexual behavior, disruption of normal nest building and maintenance, breakup of stable social groups, disappearance of alteration of normal social roles, parental desertion of young and frequently cannibalism of young. The leading question, of course, is whether or not similar processes occur in man as a result of overcrowding. There are so many variables in man which affect health and behavior, that this becomes a very difficult question. There is no doubt that some striking correlations of disease and behavior are associated with urban concentrations. Many human problems seem to be so accentuated in crowded populations that they virtually constitute an "inner city syndrome" disease, and emphysema; maternal problems such as high prenatal loss, parturitional difficulties and high infant mortlity; behaviour problems such as alcoholism, drug abuse, mental illness, and criminal assault; and social problems such as high divorce rates and social instability. More recently it has been found that high prevalence of mental illness in the lower socioeconomic classes and dense urban ghettos of several other cities. The three volume report on Manhattan entitled, "Mental Health in Metropolis," also found unusually high prevalence of mental illness in crowded inner city inhabitants.

The problem of urban man seem so strikingly similar to those effects produced in overcrowded animal populations, it is tempting to conclude that similar processes are occurring. But, as stated previously, many factors are interacting with crowding in urban manpower, malnutrition, lack of educational and recreational opportunities, various environmental hazards, and unstable social patterns—that it is virtually impossible to separate cause and effect relationships. There are enough solid exceptions to the principle of "high density produces high pathology" that one must be cautious in carrying this dogma too far. For example, the young nation of Singapore is obviously crowded—over 2,000,000 people on an island of 224 square miles, making it one of the most densely populated nations of the world—yet it enjoys remarkable good health.

The death rate is one of the lowest in the world (5.5 per 1000) and the prenatal mortality is lower than that of Britain and the United States. Major factors in this good record many be its youthful population (40 per cent of the Singapore population is under 15 years of age), and the well planned interspersion of parks and greenbelts with residential and commercial areas. One does not feel a "crowded" in generalizing as in Hong Kong, Tokyo, New York and Calcutta. Such variable demonstrate the difficulty of generalizing broadly about crowding or population density as a pathologic influence. Thus, the precise role of crowding in the "inner city syndrome" must remain a cloudily issue until more and better research studies are available.

An encouraging development of recent years is the awareness on the part of some architects, city planner anthropologist and others of the relationship between our environment and mental health. In Architectural Environment and our Mental health, Moiler explores the impact of physical space and structure on the physiological well-being of man. He noted that more buildings will be built in the next ten years than have accumulated since the beginning of civilization." And he wisely emphasized that, "We should recognize that if we continue to permit the structuring of a physical environment which is essentially hostile, to which man must somehow try to adopt, the cost to individuals and to society will be far too grat—in terms of mental illness, delinquency, poor motivation and fulfillment of our capabilities for creative work and community usefulness."

Anthropologists have also been interested in the spatial behaviour of man, and the book by Professor E.T. Hall entitled "The Hiddeen Dimension," was the first one to highlight some of the behavior patterns of man relating to social spacing. Hall showed the existence of "personal space" individuals and pointed out cultural differences in the extent and importance of personal space.

4 The Psycho-Social Consequences of Crowding

Crowding deals with a subjective psychological state when there are too many people in too little a space. The feeling of crowding is negative and has elements of stress. Typically one or many of the following characteristics are associated with crowding. Feeling of discomfort, loss of control over one's environment, lack of privacy, lack of initiative in soical interaction, fatigue, boredom etc. Crowding therefore, has perceptual affective and behavioural consequences. The outcome of such a phychological state is use of strategies that can reduce the feelings of crowding.

Some recent studies of animal populations in which overcrowding is related to various behavioural pathologies are most suggestive of the stress produced by population density. Calhoun was one of the first to note that disturbances in the territorial and ecological distribution of animals may result in behaviour pathology and physiological malfunctioning. In mice, overcrowding in laboratory cages leads to abnormal sexual behaviour, decreased reproductive and nursing capacities, aborted pregnancies, deficient maternal care for the young and disrupted nest building.

Calhoun reared a colony of Norway rats in four different pens, the rats tended to live together in one very-crowded pen. Calhoun coined the phrase "pathological togetherness" to describe this behaviour which lowered the rats' fertility and shortened their lives. Other biologists have observed that mice crowded together in cages display a marked enlargement of the adrenals, a sign of stress; and there have been reports of a high incidence of cannibalism and male impotence in crowded mice colonies, adjustive mechanisms which function to keep down population numbers when they rise above a critical point. It is not possible to dismiss these pathological reactions as unusual phenomena associated with artificial environmental conditions set up in a laboratory, for such behaviour occurs also in natural environments. Deevey has reported on the behaviour of lemmings, rat-like animals found in great aboudance in Norway.

Approximately every four years, when the population becomes excessive for the available territory, there is a mass migration in which huge numbers race blindly to the sea and drown. Such panic-like reactions to overcrowding in laboratory and natural environments are evidence of regulatory mechanisms that enable population control and, thereby, survival of the species. Whether extreme overcrowding in human populations produces analogous reactions of stress and social pathology is a moot-point. Several models of crowding and its consequences on human behaviour have been proposed and/or studied.

Stimulus Overload Model

According to this model, high population density is a stressor as it stimulates the individual, excessively. Too many people in too little space not only crowd the environment about engage in more activities creating more noise, more pollution, more distraction etc. all of which overloads the system. This results in the inability to handle all the incoming information. A consequence of this condition is an aversive feeling. To reduce the overload of stimulation caused by crowding, individuals engage in behaviors that reduce the unavoidable interference from others. At the physical level, the overload leads to fatigue and this is reduced behaviourably by withdrawing. At the cognitive level, since many cognitive stimuli, have to be handled, it results in confusion.

Some investigators also regard excessively stimulating environments as causing stress by reducing the attention span, resulting in impaired cognitive functioning. Less friendly and less helpful behaviors seen among city dwellers is an example of such strategy used for coping with stimulus overload. In homes there is greater dissatisfaction among members and desire to spend time outside is more.

Behavioural Constraints Model

High density conditions create stress by imposing restrictions on behavioural freedom. This limitation can produce a psychological reaction whereby the individual is motivated to remove the restriction by either withdrawing or by compensating for the density by engaging in behaviours that help gain control over the environment. For example in a crowded home or hostel people delineate area or seats for themselves (my chair, my corner, my window etc.) as a means of gaining some control over one's crowded environment.

Ecological Model

High density conditions lead to disruptions in accurately understanding the functions of objects in terms of their affordance. All objects in the environment, convey affordances about their functions. Eg. A fruit says eat me; a ball says kick me; a baby says protect me etc. These affordances act as cues for adapting to the environment, efficiently. High density environments block these affordances resulting in the misreading of the cues regarding their functions. For example, a family which should be perceived as supportive and emotionally positive, due to population density is seen more as a nuisance, rather than a source of joy or support. In such a perceived situation the absence of close-warm feelings among members and the desire to form smaller cliques etc. are easily understandable.

Many extended families are now breaking up into separate entities or becoming fully nuclear one of the reasons for this being crowding. The situation is further compounded by the scarcity of social roles and physical resources. For eg. The grand father plays the senior role member, even though the adult sons have taken over the functions of providers for the family. The thwarting of such roles and privileges can lead to a feeling of lack of control over one's environment. Behavioural consequences of this would be to opt for situations where control over ones environments possible, failing which, the stress is experienced without relief. This can give rise to negative feelings about the family.

Attribution Model

In this model individuals who experience lack of personal space due to crowding and therefore feel the negative attributes the stress to space encroachments. Eg. When too many guests arrive and space is short, there is sense of relief when the people leave. However, there are certain situations when high density is not very negatively perceived Eg.

An enjoyable and voluntary party. So the critical element in response to limitations of space lies in attributing the negative feelings to crowding. If the attribution is absent; then the sense of crowding is not experienced, so negatively, even though the environment is a high density one.

Stress and Arousal Model

How do high density conditions produce stress? Any physical, social or emotional stimuli that are unsettling can be described as stressful. Stressful conditions are characterized by:-

a) Stimuli that are very intense.

b) Beyond the limits of one's control.

c) Threatening, due to one's inability to cope with it.

High density environments span across these three factors making a stressful situation. Sources of stress have also been identified:

1. *Physical*: Any intense stimulus arouses stressful psycho-physiological responses. Extreme noises, temperative pain etc. are conditions that cause stress.
2. *Psychological*: Deficiencies in personality make up stress the individuality impeding the progress towards the goal. Eg. Inadequate skills, attitudes, poor leadership qualities, having an avoidant personality etc.
3. *Social/Environmental*: Social factors like collective norms traditions etc. can stress the individual by acting as barriers to goal attainment. Environmental factors that stress the individual include, noise, pollution, poor services, over crowding etc. Stress involves s situation where the individual experiences threat to his important goals and also feels that he is unable to cope with the threat. There is also an impending sense of going to be overwhelmed.

Stresses can arise from major traumatic life events like death of a spouse, parent, personal injury, retirement, pregnancy etc., or from the frequent minor hassles of everyday living. Fortunately traumatic events though extremely stressful, are rare. The minor daily stresses make up for their low intensity by their nagging frequency.

Effects of Stress on health

Stress affects the physical and psychological health of the individual. It has implications for performance in tasks as well as on social behavior. The effects of stress on physical health has been well researched. Stress is thought to be silent killer. Medical reports suggest that stress affects health to the tune of 50-70%. Stress has been implicated in the occurrence of diseases like heart attacks, hypertension, hardening of the arteries, ulcers and even diabetes. This is because stress renders the immunological system inefficient.

The more prolonged the stress is, the greater is the threat to the immune system of the body. Studies done on individuals who are in mourning or about to take an important examination have been found to show disruptions in their immune system. However, what is encouraging is, that once these high stress levels are lessened, the restoration of the immune system, is rapid. Stress inevitably arouses the organism. This is a survival oriented reaction of the organism to meet the emergency situation induced by the stress. The body's resources are then mobilized to meet the threat. This is a demanding and exhausting process. Therefore, if stress is constant or/frequent it takes a toll on the physical health and psychological well being of the individual.

Crowding qualifies as a constant stressor. In order to escape the harmful effects, the individual engages in behaviours that seek to lessen the density and thereby the experience of crowding. This in turn would lead to a lowering of the stress. Typically this manifests in the form of less helpful and less friendly behaviours. This can also explain the urban/rural divide in terms of friendliness and feelings for others. The long term effects of crowding have shown higher arousal of the system. Arousal is a form of energization of the system. The greater the stress, the greater the arousal. However, adaptations to the stress also occur. But the adaptation generally involves raising the base line. This means that the threshold for stress tolerance becomes lower. Subsequently in real terms it means that events produce less stress are also, now capable of effecting the organism.

Effects of Crowding on Social Behaviour

Since the effects of high population density are generally negative, it has consequences for social behaviour. In a study, people who were already in a crowded room were informed that more people are going to come in. The anticipation of more crowding led to feelings of discomfort. It showed up in reduced liking of even fellows waiting alongside. This is seen in a train compartment. Even as the train enters a station the possibility of more people entering produce stress. This leads to overall reduction in friendliness.

Significant reduction in helpful behaviour like making room for luggage or staying clear of the passage etc. can be explained in terms of the effects of the crowding. Individuals who are required to be in environments that are both confining and at the same time be constantly exposed to other individuals as in prisons and other settings like remand homes, free hostels etc. showed greater irritability and reduced satisfaction and harmony in social relationships.

In the community settlement the "Kiffutz" in Israel, where children are raised in a community with multiple mothering lacked privacy. This absence of privacy and constant exposure to others resulted in feelings of tension. Anxiety and aggression were some of the other characteristics that were observed.

Students living in three and four seater rooms expressed greater dissatisfaction with their roommates as compared to those living in two seater rooms. The latter also felt less control over their interactions. Some studies have shown that the grades obtained by the students in the more crowded conditions were lower than those from the more comfortable settings. These differences can be attributed to social density. Behaviours that follow, typically involve attempts to reduce the stress and unpleasantness brought about by crowding. Withdrawing and/or less friendly/helpful orientations are the most common forms of response. Table below shows the satisfactions derived from roommates in two conditions.

Satisfaction with Roommate under crowded and uncrowded conditions

	Uncrowded	Crowded
Satisfaction with roommate	4.9	3.7
Perceived cooperativeness of roommate	4.7	3.9

* Higher numbers indicate more positive responses.
(Based on data from Baron *et.al., 1976*)

The lowered frequency of interaction in high density places can be seen as the effects of less attraction for neighbours. This can then be attributed to crowding whereby lessening of contacts is a way of ensuring reducing the unpleasantness of crowding. Although it appears that high density leads to lower attraction, there is evidence that under conditions of high spatial density, this response is more characteristic of males than of females.

In an experiment male and female subjects participated in same-sex groups of six. Consistent findings on a variety of measures indicated that while males responded more negatively to group members in high than in low spatial density conditions, females, liked group members more under high density conditions. They state that while both males and females experience arousal from high density, social norms permit females to share their distress with others in their group, which leads to liking and cohesion. The same norms prohibit males from sharing stress, which causes a more negative response. In a follow-up experimental, 1976, it was found that when females were not permitted to interact with each other, their positive reactions to high density were limited.

Prosocial Behaviour. If high density leads to lower attraction and to withdrawal responses, how might it affect helping? For example, suppose you lost a contact lens or something else of value. Where would you be most confident of finding someone who would help you look for it; in a high-rise or a low-rise building? In a large city or in a small town? Interestingly, most research on how density affects prosocial behavior has been done in field settings like these. In fact, one set of studies has compared helping in building characterized by high and low density, and a second has compared prosocial acts in large cities and small towns.

In studies that explored helping as a function of building density, it was found consistently that greater density leads to less helping. Studies compared prosocial acts in high, medium, and low density dormitories. Envelopes, which were stamped and addressed, were dropped in the dormitories, and helpfulness was measured by the number that were picked up and placed in the mail. The results showed that 58 percent were mailed in the high density condition, 79 per cent in the medium density condition, and 88 per cent in the low density condition. In an interesting corroborative study, the effect of social density on compliance with a prosocial request (printed on signs) for cafeteria users to return their trays to designated areas. It was found that fewer users complied during high density periods.

The second line of research, which compared urban and rural helping, provides more equivocal evidence concerning a density-prosocial behaviour relationship. In one group of studies it was found that city dwellers were less helpful and less informative to strangers than rural dwellers. Urbanities were less likely to admit a stranger who needed to use a telephone, less likely to assist a caller who had mistakenly called a wrong number, less likely to correct over payments in making change, and less likely to report theft than rural dwellers. A number of interesting field studies have shown that resident of low-rise buildings are more willing to help others than residents of high-rise buildings.

Withdrawal. Withdrawal may be associated with high levels of social contact. Studies have found that withdrawal may function as an anticipatory response to high density, as a means of coping with ongoing high density, and as an after effect. For example, the mere expectation of high social density elicits withdrawal responses, including lower levels of eye contact, head movements away from others and maintenance of greater interpersonal distances.

Withdrawal also occurs during ongoing high density interactions: Subjects are more willing to discuss intimate topics under low density conditions, and both children and psychiatric patients interact less frequently as room density increase. Finally, it appears that withdrawal can constitute an after effect of exposure to density. Males were less likely to volunteer for another experimental session after high social density and groups of males preferred larger personal space and recalled fewer names after exposure to high density.

Aggression: If density can make people less helpful or want to withdraw; then it is possible that density can also lead to hurting behaviours. Many studies have also established greater aggression and hostility from high population areas. Apart from crowding the high density areas like the slums, require that people have to compete for scarce resources also. For eg. space, water etc. Hostilities can break out frequently in these conditions because they have conditions which are highly tension producing.

Competition is a major possible source of interpersonal conflict. This conflict can easily become inter-group conflict under conditions that threaten limited vital resources. These would be heightened by feelings of insecurity due to dissatisfaction of needs, resulting from population pressures. One study is tallisled that moderately high density conditions led to increased aggression in males, while very low and very high density conditions led to decreased aggressiveness In another study, aggression was found to increase in boys under conditions of high density. These findings can be explained in terms of competition for limited resources which males, more aggressive as a group.

Sex differences have been observed with respect to aggression. Same sex groups of eight members were placed in a room that was either large or small. Males rated themselves as more aggressive in the small rooms as compared to females. Studies of boys and girls in the class room as well as in anticipation of more cordoned, produced more aggression among males. Spatial density, rather than social density seems to have more aggressive outcomes from males.

Mean Scores on the Feeling of Crowding and Social Behaviour for Outside-Inside Density Groups

Groups		*Mean*	*df*	*t*
Outside	High density	61.00	98	49
	Low density.	60.00		
Inside	High density	89.00	48	7.49
	Low density	46.70		

**p<.01

Inside and outside conditions of population density and its effects on social behaviour were studied. It was found that density is a necessary but not a sufficient condition for producing feelings of crowding. But high density within the house had significant effects on the feeling of crowding. The following taik gives a comparison of the behaviour feelings of individual living in high low density conditions.

Social Behaviours and the Feelings of Crowding Experienced by the Residents of High and Low Inside Density Conditions

Behaviour/Feeling	*Density*	*Mean*	*t(48)*
Crowding	High	3.55	3.33*
	Low	1.82	
Privacy	High	3.07	
	Low	1.38	3.07*
Physical condition	High	3.22	
	Low	1.44	4.45*
Interaction	High	3.24	
	Low	1.38	3.20*
Neighbours	High	2.65	
	Low	1.69	3.92*
Insecurity	High	2.67	
	Low	1.24	6.5*
Illness	High	3.44	
	Low	1.71	3.39*

Withdrawal	High	3.88	
	Low	2.46	2.43*

Further, crowding affected health, feelings of security, privacy and generally produced the desire to leave. Negative moods and dislike for affiliation were also related to crowding inside the house. The feelings of crowding in high and low socio-economic studies groups have been compared, and are given below:

The Feeling of Crowding in High and Low Economic Status Groups

Dimensions of the Feelings of Crowding	*Economic Status*	*mean*	*t (48)*
Comfortable-uncomfortable	High	1.97	
	Low	2.34	2.64
Good-bad	High	1.98	
	Low	2.48	3.57*
High Congestion-low congestion	High	2.02	
	Low	2.14	0.32
Happy-sad	High	2.08	
	Low	2.48	2.85*
Pleasant-unpleasant	High	1.86	
	Low	2.40	4.00*
Positive-negative	High	1.86	
	Low	2.36	3.57*

*P< 05
**P< 01

The following table shows the feeling of crowding under high and low density conditions in terms several dimensions.

The feeling of Crowding in High and Low Density Groups

Dimensions of the Feeling of Crowding	*Density*	*Mean*	*t (48)*
Comfortable-uncomfortable	High	2.32	
	Low	2.00	2.28*
Good-Bad	High	2.40	

	Low	2.04	2.57*
High congestion-low congestion	High	2.00	
	Low	2.14	0.81
Happy-sad	High	2.38	
	Low	2.18	1.42
Pleasant-unpleasant	High	2.28	
	Low	1.98	2.14*
Positive-negative	High	2.25	
	Low	2.10	1.07

*P<05
**P<02

The Jain findings show that high density in itself does not lead to low degree of attraction towards neighbours; but high density conditions can aggravate aversive feelings if space and resources are extremely limited and have to be shared. This finding has implications for architectural and colony features that have to be kept in mind while designing high density housing localities.

In the Indian societal conditions where emotional bonds between relatives, neighbours and community have been traditionally emphasised; the western models which predict low inter personal attraction in high density areas, cannot be fully observed. In fact it is, the high socio-economic class who are also highly competitive and show less interpersonal attraction for neighbours and others, as compared to the lower socio economic group.

The typical helping out relatives, neighbours etc. in times of funerals, weddings etc. is more characteristics of the highly populated, lower economic group. The results thus show that economic factors interact with high density and societal values to moderate the effects of high density living conditions on social behaviours.

Effects of High Density on Task Performance

One of the most critical questions that can be asked about high density is whether it affects task performance. The answer has important implications for the design of all types of living and working spaces (e.g., schools, factories). Most early studies used simple tasks and were consistent in finding no decrements under high social to spatial density. As seem in the table below both high social and high spatial density led to decrements in complex maze task performance, but these decrements were more pronounced under conditions of high social density. In a field setting, also decrements in complex task performance over time in overcrowded dormitory rooms (three persons in a room built for two) were observed.

*Errors in Maze Performance as a Function of Spatial Density and social Density**

Low spatial density	34.20
High spatial density	37.44
Low social density	32.13
High social density	39.50

In another study poorer complex task performance under crowded conditions was noticed, but there was no impairment in simple task performance. Even anticipation of crowding led to poorer complex task performance. The differential effects of high density conditions on single and complex tasks can be explained using the concept of arousal.

Arousal when moderate has the properties of being a motivator. But when the organism is over aroused as in the case of overcrowding, the anxiety is experienced. Anxiety is detrimental to learning & performance. Complex tasks which need greater attention & problem solving capabilities is more adversely affected than simple tasks. So high density is more adverse for complex task performance than simple tasks. When high density conditions prevail & the expectations of task performance are also low; it yields significantly poorer functioning. In another study interaction among members was found to be an important factor that produces lowering of tasks performance in a high density condition.

Interaction is necessary for density-related task performance deficits to occur. A final and very important question is whether high density can cause *aftereffects*, as well as immediate effects, on performance. As you will recall, noise is a stressor that has been linked to consequences occurring *after* exposure, and some studies suggest that crowding has lingering effects as well. For example, it was found that subjects exposed to high density later showed less persistence at working on unsolvable puzzles than subjects exposed to low density. In an attempt to determine whether perceived control would lessen affereffects, some subjects had the option of leaving a crowded room to complete the study in a spacious setting (i.e., perceived control).

Although no one took advantage of the option, this group showed fewer aftereffects than the group that wasn't offered an opportunity to leave. The similarity of these findings to those with noise effects further suggests that noise and high density may affect people in similar ways. Performance suffering under high density, can also be explained as stress induced. Stress has been found to systematically lower performance primarily because it distracts the person from concentrating on the task on hand. Lack of concentration in turn is because of the unpleasantness of the experience. This shift in focus from task to the negative emotions reduces efficiency. Complex tasks stand to be more readily disrupted by stress as these require greater concentration and therefore suffer larger deficits.

Decision making under conditions of stress have been described as resulting in bad choices. The strategies utilised in problem-solving under stress conditions have been found to be poor. Many premature decisions are also associated with choices made under stress. Scanning of the possible alternatives for making decisions has been found to be disorganized under conditions of stress.

A tendency to close the issue on hand as quickly as possible is linked to the presence of stress. Here the individual is motivated to overcome the aversiveness of the stress and therefore quickly engages in actions to reduce it. Premature closure without considering all aspects of a given situation results, owing to the stress effects. Crowding not only induces stress and thereby lowers performance but can also be implicated in poorer decisions and premature closures. Crowding also introduces the stimulus overload condition. To reduce the overload, individuals engage in hasty and ill-conceived decisions.

Moderate levels of stress which have motivational properties of energizing the organism, has significant positive features, in terms of performance. Medium amounts of stress are considered as arousing the organism to better performance through increased mobilization of energy and better attentional focus. However, when they cross the optimal conditions, then the effects are negative and adversely affect performance. Crowding, due to its excessive stimulating features, qualifies more as a stressor and thereby has deleterious consequences on performance.

The only kind of skills that are not very adversely affected are the well established ones like driving, swimming, playing an instrument or speaking a language. Crowding as a stressor, has significant effects on the health of the individual. This in turn has negative consequences for learning & performance. Since the effects of crowding on health is also infavourable, performance can also suffer on account of reduced health in an indirect manner.

The effect of stress on learning is even more disruptive. Learning involves the acquisition of a new set of responses. This implies that the correct response exists at the 0% level of probability of occuring. This low probability response is at the bottom of the hierarchy. Learning means pushing this 0% probability response to one of 100% probability. Such an endeavour requires high degrees of attention and cognitive imputs. Stress induced by crowding, arouses the organism and such an arousal state is energizing. When learning is engaged in this energized condition, the first natural response to occur is a previously learnt 100% probability one.

The new response (to be acquired) has to now compete with the well established earlier response. By the very nature of the process, the newer response has a disadvantage and would therefore suffer, seriously.

Research evidence shows that even the mere presence of others is deleterious to learning. Crowding (which is the presence of many people) would lower learning not only because of the presence of others, but more so because of too many people. Thus, stress induced by crowding would hamper learning in an even more pronounced manner. The new responses to be learnt would in the competition with the older established ones, get depressed easily. So greater effort and attention would be required to offset this condition. Stress in such a context is clearly debilitating.

Education imparted in high density classrooms is found to be ineffective. Further, children who study in classes with large numbers, are also more likely to come from homes that are crowded, than otherwise. This would suggest that both the conditions are unfavourable for promoting learning or that extra efforts would have to be made for reasonable levels of learning to occur. In such strained conditions, students would also not readily have role models of sustained learners. To compound it, other stressors would be in operation in such disadvantaged environments. All these together pose a set of factors that are not conducive to serious learning activities. This could well be the explanation for the absence of children from the slum sections benefitting from long-term education.

Effects of Crowding on Motivation

Crowding inevitably leads to greater competition for both space as well as resources. If competition is continuous, it can produce increased tolerance for competition according to Uday Jain, who has done pioneering work on crowding in India.

Scarcity of resources typically manifests in struggle for power with a view to cornering the scant resources. In India, the immense desire for power is not so much for power itself but more because power symbolises access to resources. Obtaining railway tickets, getting admissions in educational institutions, jobs, telephone connections, licences for business and industry, etc., are all related to power as a means of need-gratification.

In Jain's study it was clearly demonstrated that high density situations lead to greater competitiveness as a way to obtaining the limited available resources. Urban people from high density areas showed more competition tolerance than those from the low density areas. Similarly, the high density urban individuals competed more than the high density rural individuals. High and low density areas in the rural setting also yielded similar results. In this situation of competing for scarce resources, people hardly find time to think or be mindful of others' needs. This can account for the apparent indifference of the present day city dweller. Given that one has limited time/energy, etc. for obtaining some benefits for oneself, expecting one to be very caring and considerate to others seems to be a very unreasonable one.

The blanket statement made about the younger generation perhaps needs to be reinterpreted using the concepts of high levels of competition generated by high levels of population. It is a clear that the earlier generation competed less for the resources as there was a larger amount that could be shared. If competition becomes fierce as the population increases, it is possibly that the individual's interpersonal relationships would be markedly affected. The biggest causality of such a condition would be consideration for others.

Studies show that there are urban/rural differences in the extent to which they tolerate competition. Greater tolerance for competition has also been associated with the need patterns of the high and the low competitive individuals. Low competitiveness is associated with helpful and considerate behaviours while high competitiveness is related to order, neatness, organization, etc. The latters needs are likely to be more readily disrupted in a crowded environment.

The reduced consideration and helping behavior found among the high competitive individuals can be easily attributed to their negative perceptions of others who are viewed as rivals. This is due to the fact that too many persons would not only increase the competition but also intrude into their neat and orderly styles of functioning. Such a situation would naturally yield less consideration and helpful behaviour. High density dwellers have to encounter vast numbers in their daily activities. They try to conserve their psychic energy by keeping interaction to the barest possible number of people. In such a context, friendly, helpful and considerate social interactions are highly unlikely. This is more clear among the more educated who are more competitive and thus uninterested in nurturing supportive behaviours with others with any degree of spontaneity.

So, the explorations of several density-behaviour relationships shows that high density has various kinds of negative effects. High density is like any other potential stressor and it has !

(1) Immediate effects, such as physiological arousal and negative states.
(2) Coping responses like withdrawal, aggression etc.
(3) After-effects and cumulative effects like health and performance.
(4) Motivational consequences (increased tolerance for competition)

Summary of Theoretical Perspectives on Crowding

Conceptual Approach	*Critical Causes(s) Crowding*	*of Primary Coping Mechanisms*
Social overload	Excessive social contact too much social stimulation	Eacape stimulation; prioritize input and disregard low priorities withdrawal
Behaviour constraint	Reduced behavioural freedom	Aggressive behaviour leave situtlon coordinate actions with other
Unwanted interaction	Excessive unregulable or unwanted contact with others	Witdrawal; organization of small primary groups
Interference	Disruption or	

	blocking of goal-directed behaviour	Create structure aggression; escape
Ecological	Scarcity of resources of outsiders	Defense of group boundaries; exclusion

All these models are overlapping and are supplementary in terms of explaining the stress experienced crowding and the behavioural consequences that follow, being stress-alleviating in nature.

Recent conceptualisations on the effects of high density revolve around two concepts control and privacy. Perceived control is regarded as a potent mediator of stress. When we believe that we can control a stressor or other aspects of a situation, the aversiveness of stress appears to be reduced. It follows that even if no other problems are apparent, losing or not having control can also be stressful.

Several researchers have propsed that high density can cause a loss of control (or prevent some one from ever having control) and that this loss of control is the primary mechanism by which density causes stress. Can we really explain the negative effects of high density as a loss of control? The answer is "Yes," even though at times the explanation seems a little strained. Some explanations fit well the *notion of unwanted interaction* is a control-based perspective. Negative effects are the result of contact that is too frequent which makes control over when, where, and with whom people interact difficult to manintain.

As a result, interactions become unpredictable and frequently unwanted. The *behaviour constraint* notion is also control-based, decribing high density as eliminating behavioural options and reducing freedom to behave as one might like. For example, having inadequate space can constrain our behaviour by making it impossible to control the nature of interaction with others. We can generally control the degree of intimacy in one-on-one interactions by adjusting the distance we stand from people but in a very crowded room we may find ourselves with no choice-we must stand close to people whether we know them well or not. In this manner, we can lose control over intimacy regulation.

Interference can also be viewed as a threat to control as our attempts to achieve one or another goal are repeatedly blocked or disrupted. And resource problems the focus of ecological models, can limit our choices and restrict our ability to exercise control. To some extent, all of these consequences of high density that have been related to negative effects do cause a reduction in control. But, is there any direct research evidence for the relation between density and control? Research examining the links between high density and loss of control has taken two different tracks. The first has been to manipulate personal control (provide some subjects and not others with perceived cantrol) and see if it has any effect on experience in high density settings. Example subjects response to riding in an elevator crowded with confederate of the experimenter. They observed a tendency for people to gravitate toward the elevator floor selection panel (a control panel, if you will), so they manipulated whether subjects were able to stand near it (high control condition) or not (low control condition).

The results indicated that subjects standing near the panel felt better than those in a defferent corner. They also thought that the elevator was larger than subjects not near the panel. A some what different approach found that noise was associated with negative after effects which was reduced when subjects had some degree of control over the noise. A similar, study with density instead of noise was carried out. Once again, negative after effects were associated with stress when control was not available. Subjects who had more perceived control, on the other hand, did not exhibit negative after effects following exposure to high density. In a high density condition in the laboratory, subjects had varying degrees of control over the setting. Those with control felt better than those without. These studies demonstrate that control can reduce the effects of high density relative to no-control, high density experiences. They do not, however, demonstrate that high density itself has consequences similar to those associated with loss or lack of control.

However studies conducted in field settings found that chronically high residential density was associated with helplessness-like behaviour. Learned helplessness is a syndrome in which people who are exposed to uncontrollable setting learn that they cannot control the setting, and hence stop trying to do so. This manifests itself in reduced motivtion and congnitive activity. Subjects, who were children and adolescents showed symptoms of helplessness that were associated with density in their homes.

Symptoms of helplessness were also found among people exposed to high density in their residential environment over prolonged periods. More recently this has been directly linked to perceptions of control in residential settings. As people relinquished their beliefs that they could control these settings, their behaviour became increasingly like that asssociated with helplessness. All of these studies taken together suggest that control is involved in negative high density effects. When control is available, high density has less impact on pepole than when it is not available. Further chronic exposure to high density appears to be asociated with learned helplessness.

Crowding in the Home and in the Schools

Imagine yourself growing up in a small apartment with five other people who are continuously interacting with each other and with you. With so many people in so little space, you may grow up to feel the world is a complex place in which you have little power to influence events (What consequences does this have? When we come to believe we cannot control our outcomes by responding appropriately (as may result from living in high density conditions), we no longer perform effectively in a number of situations. This syndrome is called "learned helplessness." Two interesting and provocative studies demonstrated the relationship between residential density and susceptibility to "learned helplessness".

In one experiment it was found that children who lived in high density conditions were less likely than those in low density to try to control the administration of rewards they were to receive. In a second experiment subjects from both groups to an initial frustration task on which responses and outcomes were noncontingent children from high density homes did significantly worse on a subsequent task for which outcomes were contingent. Thus it appears that home density is an important determinant of both the use of control and performance after frustrating noncontingent reward situations. Besides the home environment what other settings may be source of 'learned helplessness' training?

One important culprit may be the schools. As classroom size increases, learned helplessness training begins to occur. They argue that larger classroom sizes lead to lower student expectations for control of reinforcement, because teacher feedback concerning student work becomes less descriminative. For example, as class size goes up, individualized student-teacher interactions decrease, and generalized (rather than individualized) praise and criticism increase. Clearly such conditions could lead to a state of learned helplessness with its negative consequences for performance. This could have profound consequences for learning. An alternative to the density-intensity model has been proposed by Freedman. He agrues that density only intensifies reactions that would occur in any case in a particular situation.

High density according to him, heightens the importance of other people and magnifies our reactions to them. Thus high density would intensify the aversiveness of an unpleasant situation and intensify the pleasantness of positive situations. From this point of view, a number of factors can lead to a negative reaction in a high density situation. One of the benefits of such a model which utilises the notion of intensfication; is the link between crowding and contagion.

Contagion is the rapid spread of behaviours or emotions from one person to the group. Eg. People coming back after viewing a game. The negative effect, (if the home team loses) would be quickly passed on among all those returning. This model provides a frame work that specifies three factors as important in determining the experience of crowding. They are:

(1) Individual difference

(2) Situational conditions

(3) Social Conditions

Identifying individual difference variables can help determine who among the selected persons will be more sensitive to the constraint of space. Studies have already established gender differences in the aversire feelings, due to crowding.

Some persons prefer to sit for away from others while others prefer to sit closer. This can also affect the degree to which crowding is experienced.

Personality characteristics also moderate our reactions to high density. It has been found that internals (who feel they control their fate) display a higher threshold of crowding than externals (who feel events are controlled by outside forces). The primary problems encountered in high density settings may also affect the ways in which individual differences are important.

People who screen themselves from interaction and organize their surroundings were better able to cope with social density than were people who did not screen themselves. One would expect that this variable would be less important when spatial problems were paramount. A final individual difference variable that has been linked to reactions to high density is one's adaptation level (from past experience under high density conditions). A number of investigators have hypothesized that people with a history of high density living are less likely to experience crowding in a novel situation than are people with a history of isolation. For example it has been found that the Japanese, residents of HongKong and the Logoli (all of whom live under extremely high density) have developed social mechanisms that may be viewed as adaptive for high density living.

In addition to being moderated by individual difference variables, our reactions to high density are also affected by *situational conditions*. An important situational condition is the degree of control we have. Complementing those discussed earlier, several additional studies have found that allowing people increased control over a situation, leads to less perceived crowding. The applied potential for introducing control in high density conditions, is phenomenal. For e.g. Individuals living in high denity situations, can be given training for enhancing control. ie. how to share space and ensure privacy. This is bound to help alleviate crowding and it negative effects. Other situational factors as whether we are in a primary environment (like a home) or a secondary environment (like a restaurant), and the extent to which other stressors, such as noise etc. are involved, also affect crowding. When we are in a primary environment we may be more likely to experience crowding.

Finally, in terms of activities there is suggestive evidence that we are more likely to experience crowding when engaged in work, rather than in recreation. Life styles that increase social contacts like picnics, games, travel etc are more likely to generate feelings of crowding in dense environments than life styles that decrease the likelihood of social contacts. In addition to individual differences and situational conitions, *social conditions* can be manipiulated to affect whether or not we are crowded. These are variables that make up the social "climate" of a high density situation such as the degree of friendship and the level of social interference and interaction. For example our relationship with the people were with, may determine how crowded we feel: Less crowding is experienced with liked rather than disliked others and with acquaintances than with strangers.

To the extent that we experience social interference (interruptions) by others or excessive proximity or intimacy (too direct eye contact or body orientation) crowding is more likely. Finally crowding is more often experienced in unstructured task situations than in structured ones. High density within primary groups (e.g. families) has repeatedly shown fewer efects than it has in other groups—when density *within* apartment units in examined, for example it appears to be negligible as a factor associated with illness or behavioral difficulties. Degree of acquaintance and relative position of an individual in a group's dominance hierarchy also affects crowding.

The presence of friends or the possession of high status tends to blunt the aversiveness of large numbers of people or cramped spaces. Since some of the conditions and factors that mediate the experience of crowding have been delineated; it is worth considering how one can modify existing environments or plan new ones to mitigate the effects of crowding.

Some architectural modifications

To begin with, greater ceiling height is asssociated with a feeling of less crowding. Rectangular rooms seems to elicit less crowding than square rooms, of the same area. Rooms that have visual escapes such as doors or windows are rated less crowded than rooms that do not have such escapes. The adding of flexible partitions in rooms, lessens feelings of crowding. Increasing the brightness of rooms by providing suitable wall colors leads to less perceived crowding. The presence of visual distractors e.g. pictures, some movement etc. also helps reduce crowding. Finally, studies that have evaluated the aversive effects of Social *vs* Spatial density have established that social density is more negative than spatial density. Architects and designers must address themselves to the fact that too many people produces greater sense of crowding than too little space.

Architectural design for crime prevention

Cities are the hotled of crimes of different kinds owing to a peculiar mix of people, social, physical and housing factors. Large scale rural to urban migration; the concentration of the underprivileged in core unban areas, crowding of population into higher and higher densities all significantly affect life styles environmental conditions and crime rates, in the large metropolitan areas. It is now becoming increasingly clear that the physical and architectural designs of residential environments can elicit attitudes and behaviour from residents which contribute in a major way towards insuring their security Residential environments with certain physical features have crime-inhibiting qualities.

As a starter, 'semi-private' spaces should be 'designed in' to facilitate social interaction between residents and thus encourage the development of social order. Such areas should exist both inside and outside the building and should be planned so as to make pleasant social interactions inevitable. This is specially true of many South Asian countries and particularly India, where dependence and interdependence rather than independence is a valued social goal in interpersonal relationships.

Many homes have extended families, which involve the living together of many individuals of different generations. So our housing colonies shouid be totally designed to suit our socio-cultural environmental therefore needs. So houses that alienate cannot be appropriate for India. Yet, colonies that grow up where spaces are 'nobodys' can become dumping grounds easy for encroachments or neglect and subtly support crime and randalism. In order to take care of this situation the design of a 'defensible space' has been proposed. This notion implies that building should be so designed that all the space appears to 'belong' to some individual or group. Such areas can be easily depended and therefore promote surveillance and thus reduce unauthorised activities and crime.

In practice this means that areas around buildings e.g. service lanes, parks playgrounds etc. must belong to the building. Then these areas would be perceived by both the residents of the building and outsiders as extensions of the building. The otherside of this issue is that public areas such as hallways, corridors, stairways etc. should be eliminated as far as possible. This can be done through horizontal rather than vertical constructions. Where this is not possible then open 'air spaces' like courtyards combined porches or hallways that are accessible to only a small group of residents should be constructed. These designs would facilitate territorial functioning. If corridors have to be provided; they can be made to run along the window side. This is easy to be monitored.

Designs should also help police surveillance. Often buildings are not open to police viewing or even reaching at times of need. All residential housing units should have a garbage dump, a community centre and a self monitoring cell. It should be the responsibility of a duly elected group of representatives to keep their environment clean, green and protected. A national and regional regulation regarding acceptable levels of density should be evolved and stringently enforced. To insure lower population density, attitudinal and behavioural change at the individual level should be taken by the government on warfooting in a massive scale.

5 Community Ecology

Communities may be distinguished as major or minor. Major communities are those which, together with their habitats, form more or less complete and self sustaining units or ecosystems, expect for the indispensable input of solar energy. Minor communities, often called societies, are secondary aggregations within a major community and are not, therefore, completely independent units as far as circulation of energy is concerned. The biotic community along with its habitat is called an ecosystem the term "ecosystem" has been loosely applied to units of various sizes and characteri-stics but had best be limited to distinctive combinations of air, soil and water conditions with vegetative, animal and microbic life that possess functional unity.

When a number of species come together to form communities each fits in a different niche and plays a different role in the internal dynamics of the community. Dominance is the relative control exerted by organisms over the species composition of the community. Species exerting this important control are called dominants. Plants are more frequently dominant in terrestrial communities than are animals.

In aquatic communities, animals are relatively more important in this role. Although dominance is often not developed. Dominance is most commonly expressed in the reaction of an organism on its habitat. Dominants shoulder the full impact of the climate or the environment but modify this effect for other organisms within the community by tempering light, moisture, space, and other conditions. Only those organisms that find these modified physical conditions tolerable can exist within the community.

Dominants are ordinarily the most prominent species in the community, make up its greatest mass of living material, and serve as the major source of food, substrate, and shelter for the animals that are present. In a forest community, trees are dominant. They decrease light intensity increase the relative humidity intercept precipitation, monopolize most of the moisture and nutrients in the soil, decrease wind velocity, and furnish shelter and food for animals. Grasses play a similar, though, less conspicuous, role in prairie communities; sedges, rushes, and cattails in marsh communities, sagebrush in the arid habitat of the Great basin, mussels and barnacles on a rocky seashore; and so forth.

Sometimes dominance is demonstrated in coactions, direct effects organisms on each other. In some fresh-water ponds, carp and suckers may consume much of the submerged vegetation. This coaction thus prevents the plant constituents from assuming their usual role in the community, and by so much prevents the occurrence of animal species that depend directly upon the plants. These fish also react upon the habitat by stirring up the bottom, from which they derive organic matter, thereby greatly increasing the turbidity of the water.

Penetration of light into the water is reduced, greatly handicapping sunfish, bass, and other species which locate food visually. In primeval days, bison our western great plains fed on the luxurient taller grasses more extensively than on the short grasses, with the consequence that, over extensive areas, short grass species replaced tall grasses almost entirely. Thus bison were coactant with and dominant over the composition and character of the community. In a similar manner, when, European meadow voles are numerous, they reduce the vigour and prevalence of the grass dominants in consequence of their feeding and tunnelling in the ground, and flowering plants which are normally absent or scarce appear.

Over-populations of the European rabbit alter the character of the forest by feeding upon the seedlings of oak, beech, and hornbeam to the exclusion of other species. When introduced into Australia, the European rabbit converted grassy areas into desert-like tract. Although animals are more common coactors than plants, plant pathogens may occasionally exert dominance in this way. As a notable example, chestnut tree blight (a fungus) virtually eliminated the chestnut tree from the deciduous forest of eastern North America during the first few decades of the twentieth century. This fungus infects the cambium, forms pustules under the bark, and causes the bark to fall off and the leaves to wilt.

The consequent opening up of the canopy with the death of the trees has allowed the extensive invasion of new species of shrubs, herbs, and animal inhabitants. Trees are the dominants in a forest community, but species in the lower stratum of shrubs modify the habitat still further and even the herbs exert some control over the physical conditions on the surface of the ground. A subdominant species must tolerate the conditions established for it by the dominants; but it in turn is a modifier of the community composition in a secondary manner.

By influence is meant effect upon the abundance, health, and activities of other organisms but not to the extent of directly excluding species from the community.

Influence is conspicuously expressed through coactions, but it may be effected through reactions as well. Insects many partially or wholly defoliate a tree; a pack of wolves may diminish a population of deer over winter; squirrels may bury acorns and nuts and thereby aid germination of them; parasitic or poisonous plants may lower the vigour or destroy the life of some other plants or animals; animals may burrow into the soil and thereby increase percolation of water and air, a benefit to plants; and all organisms, upon death, add organic matter to the habitat. These and other actions influence the community, but unless these influences become extreme they do not determine whether or not other species will occur in the community. Influence, then is of essentially the same nature as dominance but is less vigorous in the modifying role that it plays.

Evaluating and Classifying Animals Ecologically

One the most important, yet difficult, problems in ecology is determination of importance values of organisms to the structure of communities and of their roles in community dynamics. Various statistical criteria have been used, but we will be concerned only with some of the more general evaluations. A basis for classifying species is exclusive when it occurs only in a single area, habitat, or community; characteristic (selective, preferential) when it is abundant in one area or community but also occurs in small numbers elsewhere; or ubiquitous (indifferent) when it is found more or less evenly distributed in a wide variety of communities. The terms given in parentheses are synonyms used by plant ecologists.

Exclusive species are often rare and of little importance in the dynamics of a community, but when they are conspicuous they often make useful indicator species for identifying and recognizing community units.

The recognition of characteristic species presents special difficulties since one must decide how much more abundant a species needs to be in one community to be sure a definite preference over another is indicated.

Another criterion for evaluating species is by their density or numbers of individuals present per unit area. Other things being equal, a species in time of high population affects other organisms to a much greater extent than it does at times of low population. A species that is permanently more abundant than another will consume more food, occupy more reproductive-sites, and demand more space; hence its influence will be greater. Predominants are the more numerous constituents of a community, in contrast to members, which are species of lesser importance. The dividing line between these two categories is an arbitrary one. The time and duration of occurrence of species in a community affect the amount of influence it exerts. Generally, the longer the yearly period during which a species is active, the more important its role becomes.

Species may be classified on a temporal basis into perennials, those which are active in a community throughout the year, year after year seasonals, which are present or active only part of the year; and cyclics, which may be very important some years but of negligible importance other years, as evidenced by their wide fluctuations in numbers. Even though present, a species is usually considered inactive when it is hibernating or dormant or when it is represented only by eggs, spores, or encysted stages of its life cycle.

The effect produced on the community by individuals and species may be modified by the way they form secondary groupings within the community. These minor aggregations of plants and animals are called societies and are of various sorts. Layer societies occupy different strata, such as the subterranean, ground, herb, shrub, and tree societies in a forest; local societies are usually parts of layer societies but are more confined in area, as groups of animals occupying an ant hill, a rotting log or stump, or a restricted but distinctive area of ground; and seasonal societies include all the organisms at particular times of the year. Other factors that affect the influences of a species in the community are the size of individuals, their metabolism, food habits, and general behaviour.

A moose consumes more food than a mouse, and a warm-blooded mouse more than a cold-blooded salamander of the same size. A carnivore at the top of several food-chains affects the lives of more different species in the community web of life than a herbivore feeding on plants. Borrowing rodents react on the habitat more than do most birds. One factor may cancel another. An individual of a perennial species of carnivorous mammal certainly eats more than an individual cold-blooded herbivorous insect of small size and active only during the warm season. Yet there may be 1000 insects to one mammal, so that in the aggregate single species of inset may actually produce more disturbance than a single species of mammal.

The difficulty of evaluating the relative effects of species is partly alleviated by calculating their respective biomasses and amounts of energy used. The biomass of a species is the average weight or volume of individual, multiplied by the total number of individuals present. The computation of the biomass of each species thus corrects for differences between species in size and number of individuals. Because of differences between species in body moisture and amount of inert substances such as endoskeleton, chitin, shell, and the like, biomass is expressed with greater accuracy in terms of dry weight than wet weight or volume, and is even more accurate if given in terms of carbon or nitrogen content, or calories. Attempts have recently been made to compute more significant biomasses of bird populations, using physiological constants. A biomass composed of few but large individuals has a lower metabolic activity than an equal biomass composed of a large number of small individuals.

Productivity

A characteristic of communities that has become of considerable importance in modern ecological research is productivity. The number of individuals or biomass in a community at any one time is the standing crop. At the beginning of the year or reproductive season the standing crop is usually small, but as reproduction and growth take place there is an increase in the amount of organic matter making up the biomass of the community.

The production of organic matter per unit of time and area is productivity. Productivity is thus a rate and is commonly indicated on a yearly basis, but it is also possible to measure monthly, weekly, or daily production, small standing crops may have a high productivity and large standing crops a low productivity, hence biomass alone does not give a measure of productivity. The largest standing crop which a habitat can support without deterioration, or the maximum number or biomass of animals that can survive the least favourable yet tolerable environmental conditions during a stated period of time, is the carrying capacity.

Carrying capacity is determined not just by the amount of food available, but also by shelter, social tolerance, and other factors. A variety of methods are being used to measure productivity of different kinds of organisms and of different habitats. It is desirable to indicate productivity as accurately as possible in descriptions or analyses of community dynamics.

Succession

Communities are in a more or less continual process of change. These changes result in part from the reactions and coactions of the organisms themselves and in part from such external forces as changing physiography, changing climate, and organic evolution. The habitat is usually affected as well as the community, and as the habitat changes, new species invade it and become established, and old species disappear. These changes are especially noticeable in dominant species, since these species exert a controlling role over the composition and structure of the community as a whole.

The replacement of one community or ecosystem by another is succession, and succession continues until a climax or final stage is reached. Succession is a process. The series of steps or communities comprising a successional sequence leading to the climax is the sere. Seres are sometimes classified according to the predominant force that is bringing them about. These forces are biotic, climatic physiographic, and geologic and their resultant seres are commonly called bioseres, cliseres, eoseres, and geoseres.

Succession that begins on a sterile area where conditions of existence are not at first favourable—as, for example a newly exposed sand dune or a recent lava flow—is termed primary succession. The term secondary succession refers to community development on sites previously occupied by well developed communities or succession on sites where nutrients and conditions of existence are already favourable such as abandoned croplands, plowed grasslands, cutover forests or new ponds. As would be expected rate of change is much more rapid and time required for completion of the sere is much shorter in secondary succession. Finally it is important to distinguish between what may be called autotrophic succession and heterotrophic succession. The former is widespread type in nature that begins in a predominantly inorganic environment and is characterized by early and continued dominace by autotrophic organisms.

Heterotrophic succession characterized by early dominance by heterotrophs occurs in the special case where the environment is predominantly organic as for example in a stream heavily polluted with sewage or, on a smaller scale, in a fallen log. Energy is maximum at the beginning and declines as succession occurs unless additional organic matter is imported or until an autotrophic type but is usually maintained or increased during succession.

Biotic Succession

Biotic succession is brought about by forces inherent within the community and in the activities of the plants and animals themselves. The most important of these activities are the organismal reactions and coactions that produce modifications in the habitat and interrelations between species. Important reactions involve filling in of ponds with plant and animal remains, the addition of organic nutrients to sterile soil, and the reduction in light intensity by increasing density of plant growth.

With progressive improvement of the soil and changing light and moisture conditions, a series of new dominants come into the area. When invasion of new species occurs, intense competition develops; if the invaders are successful, the old species disappear as a new community replaces the old one. Contributing factors that may be involved are differences in growth and dispersal rates, which are different for different species. After a forest fire or logging operation, herbaceous plant growth is immediately stimulated; since herbs grow rapidly, they become dominant within a year or two. Shrubs begin to spread, and tree suckers or seedlings also appear quickly, but because they require a longer time to reach maturity, several years may elapse before they gain control of the area. Succession is considerably influenced by the kinds of propagules available in the vicinity. Seeds, spores, and the like are dispersed more or less readily, depending on form.

Some kinds of animals roam more widely or spread more readily into new areas than do other kinds. The composition of the community that develops and the rapidity with which it becomes established depend, in large part on the rates at which different species invade. If a forest is cut, and the land cleared to the soil, a succession of plants will invade and grow on the exposed surface. The first plants which invade are typically those capable of seeding in quickly on disturbed land or capable or generating from viable roots left within the soil. This depends not only upon the former vegetation, but also upon seeds from surrounding plant communities, as well as the characteristics of the soil and climate.

Typical invaders in the United States are herbaceous weeds such as ragweed (Ambrosia sp.), lambs quarters (Chenopodium album) T dock (Rumex sp.), horse weed (Erigeron canadensis), plantain (Plantago sp.), crab grass (Digitaria sp. et al.) and woody vines such as honeysuckle (Lonicera sp.) and poison ivy (Rhus taxicodendron). As soon as the first plant germinantes and becomes established, the community increases vastly in physical complexity. A surface of bare earth now has shaded areas which differ in light, moisture and humidity. This enables other seeds to grow, and new species become established. Woody plants appear, again those capable of withstanding disturbed environments and harsh conditions, such as biack locust (Rohbinia sp.), sumac (Rhus sp.) and hawthorn (Crataegus sp.), Pioneer animals also arrive, including ants, beetles and flying insects. Birds begin to forage over the new community searching for seeds and insects. Small mammals may venture forth from adjacent forests or grasslands.

Lack of animal traversing the area adds organic nutrients. Within a year or two a surprisingly complex community has arisen, and is growing rapidly. As plants grow, they create more shade, and thus continuously modify the light and moisture conditions on the surface of the ground. These initial pioneers which thrived on total sun exposure are now less favoured and more shade-tolerant seeds can germinate more successfully. Woody shrubs emerge above the herbaceous layer and compete more successfully for the ambient light. The shade developing beneath the herbaceous and shruby layers permits the growth of other tree species such as maples, elms or oaks which could not tolerate the initial exposure of sun and wind. This represents the first beginning of the forest.

Further succession continues to involve all the dynamic processes of plant and animal competition. Over a period of years, the initial invaders and pioneers begin to drop out. Trees mature first the fast growing locust and sumac, but as they reach the limits of their growth and their relatively short life-spans the slower growing oaks (Quercus sp.), maples (Acer sp.), and elms (Ulmus sp.), begin to take over. The forest may now be 40 to 50 years old, reaching a height of 50 to 60 feet.

The ground cover and intermediate layers have now matured to include those shade-tolerant species capable of living in the forest under study. The insect and bird populations have changed from open field forms to forest species. By 70 to 80 years, the forest may be approaching relative maturity, with tall trees in the 60 to 80 foot range, deep shade and cool moist conditions on the forest floor capable of suporting the rich array of life. The climax first may be approached in 80 to 100 years, when it becomes a stable and self-perpetuating community. Processes of change still occur.

Old forest trees die and crash to the ground, creating a microcosm of renewed successional development, but the main forest continues in equilibrium as a permanent community as long as environmental conditions remain favourable. If it is cut, the process of succession begins again. If it is defoliated, flooded, burned or dehydrated, it will change and attempt to seek a new form of equilibrium with its environment. The tree species that finally emerge as the dominant—members of the climax community vary according to the climate, topography and soil type. It would be incorrect to assert that the process of succession is always consistent and predictable. It may take a variety of forms, and may become arrested at any of several stages.

Early in the shrub stage, the biotic community may become overwhelmed by excessive growth of woody vines such as honeysuckle or poison ivy, so that trees cannot develop adequately, or it may stay in the herb or shrub stage if the land is subjected to continuing disturbance by fire, pollution, or land scarring. Throughout the countryside, especially around cities, one can find many examples of disturbed and unattractive vegetation that never gets out of the weed, shrub or vine stage.

The above discussion has mainly concerned a deciduous forest community, but a similar account could be given of coniferous forests, and, in fact, of all biotic communities. Thus, a prairie grassland, a tropical savannah, a desert, and estuarine marsh, a cedarbog, a pond, a lake, a delta, a mangrove swamp, or a coral reef all have characteristic patterns of development and succession. In some communities, successional development may involve hundreds or even thousands of years before mature climax conditions are achieved. Primary tropical rain forests, for example, are the culmination of thousands of years of succession development. Once cut, they are not recreated in a few decades or a hundred years. It should be emphasized that succession is also a naturally occurring process and is not confined to disturbed habitats.

Theoretically, each biotic community tends toward maturation and stable climax conditions, but there is sufficient natural change in ecosystems so that succession is a continually recurring process in many situations. The grassland and forest are subjected to natural fire as an ecologic process, the desert to flash flooding, the marsh, bog, pond, lake, delta and swamp to filling, and the coral reef to the continual processes of construction and destruction. Human activities often increase and accelerate these processes far beyond the natural rates of occurrences. In doing so, man impedes the development of maturity and ecologic stability in. biotic communities.

An example of natural succession is the filling of a small pond and its eventual conversion into a terrestrial community. As the pond ages, it becomes more filled with aquatic vegetation and silt. Emergent vegetation surrounds the edge creating a marsh community. Accumulations of organic material and sedimentary deposits fill in the water volume, soil is formed at the edge, and terrestrial grasses an hebs become established. Eventually, shrubs and trees complete the process.

Man accelerates this process by increasing the siltation process and enriching the waters with nutrients so that more extensive plant growth occurs. The process can be retarded by dredging the pond or manipulating the water level to reduce aquatic vegetation. Lakes proceed through some of the same stages, but if sufficiently deep, a point is reached in lakes where the scouring action of currents slows or even balances the situation and sedimentation rate. Another example is the volcanic island of Krakatoa in the East Indies which blew up in 1883 and virtually destroyed all plant and animal life on it on two adjacent islands. In 1886, it was apparent that the pioneer plant stage consisted of a crust of blue-green algae covering the lava.

A few mosses were also present, as were many ferns, and a scattering of some 15 species of flowering plants including 4 species of grass. The vegetation stage following the algae consisted predominantly of ferns, but the grasses had become dominant over most of the island. By 1906, woodland had appeared, which has since developed into an increasingly luxuriant mixed forest. Dispersal of seeds, spores, and other propagules was effected by wind, sea, animals, and man, in that descending order of importance.

The order in which the propagules reached the islands greatly influenced the succession that occurred; as vegetation developed it reacted on the soil and habitat, bringing about conditions amenable to the return of the tropical rain forest. One year after the explosion, the only sign of animal life on the island was a single spider, six years after the explosion flies, bugs, beetles, butterflies, and lizards were also reported. In 1908, 202 species were found on Krakatoa and 29 on a nearby island.

There were no earthworms, snakes, or mammals present, but there were many spiders and centipedes, a number of insects, 2 species each of land snails and lizards, and 16 species of birds. Bats and earthworms were found in 1919. A more through survey in 1921 revealed 770 species of animals, including rats, apparently introduced in 1918. In 1933, 1100 species were found, but true forest mammals had not yet appeared and many families of forest birds were unrepresented. As with plants, the invasion by animals depended on wind, sea, a man and other animals, in that descending order of importance. Survival and establishment of the animals species were correlated with the stage of vegetation that was reached. It is of interest, however, that scavenger species appeared first, then omnivores, herbivores, and finally predators and parasites.

Succession of animals depended in large part on the speed with which they reached the island and on their finding proper food and shelter. Bioseres may be broadly grouped as prisers and subseres, depending on whether they develop on primary or secondary bare areas. A primary bare area is a sterile habitat, such as rock, sand clay, or water. A secondary bare area is a denudation resulting from temporary flooding, fire, logging, cultivation, overgrazing, or other phenomenon that does not produce an extreme disturbance of the soil or substratum. Since in the latter the habitat has already supported community life, and since the soil or substratum is already in an advanced stage of development, the resulting subsere progresses rapidly and the early pioneer stages of the prisere are not usually required.

A subsere will develop following the destruction of the stages in the subsere will be influenced by the particular priseral community that was destroyed. The early stages or communities that make up the prisere depend largely on type of bare area on which the prisers originates. As succession proceeds, however, later stages in the various seres in any area having a relatively uniform and humid climate come to be more and more alike. The successional development from widely diversified communities in initially different habitats to closely similar or identical climax communities in habitats that have also become much alike is called convergence. Biotic succession proceeds rapidly and appears to reach a final, permanent stage in just a few decades or centuries.

The climax stage of the biosere is undoubtedly more nearly stabilized, self-maintaining, and in steady state in its particular habitat than are the seral stages, yet it also is subject to gradual change over long periods of time. The climax, as well as the seral stages, changes with climate, physiographic forces, and evolutionary processes. However, the clisere usually requires a few thousands of years before changes in the community structure or composition become evident. The progress of an eosere is even slower, that of the geosere, slowest of all. The climax, therefore, is best defined as thee last obvious stage in the biosere.

The climax may be recognized by the fact that in a uniform climatic area all seres tend to coverage into it, and by its steady state in respect to structure, species composition, and productivity. In the climax community, all species, including the dominant species, are continually able to reproduce successfully, and there is no evidence that new and different species are invading. In seral communites, on the other hand, the developing new growth, particularly evident in the dominants, contains many individuals of invading species which will eventually take over and replace species already present. It is thus important for man to understand successsional processes so that he can manage biotic communities for maximum productivity and aesthetic concerns.

In agricultural lands and harvested forests, man has learned to maximize productivity for his own economic gain. This is done by artificially maintaining communities in early successional stages, where productivity is greatest, and by channelling all possible production into the species to be harvested. That is, man maintains a cornfield or a wheatfield in an early productive stage by excluding all other pioneer species. He maintains a forest in an early successional stage and directs all productivity into the species of tree he wishes to harvest. This is all well and good provided the cost of such management is not too high in terms of toxic chemicals that must be applied to eliminate other species, or in terms of chemical fertilizers that must be applied to support an artificial single-species community.

Quite another principle prevails in the management of parks and sanctuaries where man wishes to maintain balanced plant and animal communities for a esthetic and scientific values. Here it becomes essential to understand and promote natural successsion so that mature and stable communities are achieved. In this day of economic pressure it is increasingly difficult for man to appreciate that value of natural biotic communities. What good is a mature forest if it is not harvested, what good is an estuarine marsh, what good is a relic prairie? These are all questions being asked in greater frequency by citizens focusing on economic progress. They challenge the ecologist to explain the message of ecology in factual terms, and to emphasize the importance of balanced biotic communities to the total life support capabilities of the world.

The answers are not easy, and are not always available. Much remains to be learned about the full range of functions of an estuarine marsh, for example, in maintaining, the productivity of the oceans. This represents one of the greatest current challenges for ecology; namely to understand and explain in a meaningful way without great fanfare and unreasonable emotion, the importance and value of diverse biotic communities. Otherwise economic man will see little reason why he cannot continue to covert the entire world to paved streets, grass lawns, grain fields and greenhouse.

Hydrosere

Ponds is a popular terms for lakes of the third order that are small, shallow, and when mature, have rooted vegetation over the bottom. There is no clear distinction between the ponds and lakes of the first and second orders. The littoral zone of eutrophic lakes, for instance, is pond like in habitat and organisms. Floating and emergent vegetation commoly occurs around the margins of ponds to form extensive tracts of marsh, the pond habitat may originate as shallow basin, as a large pool in a stream as the result of the filling in a lake, or form a stream dammed by beaveas, man, or landslide.

Slow-flowing rivers are esssentially elongated ponds, and have a similar fauna. Because of the slight water movement in ponds, the surface film becomes an important microhabitat for some species, pond water temperature is often uniform at all depths, but during warm sunny weather, ponds well protected from the wind may show considerable stratification, not only in temperature, but also oxygen content and other characteristics. Daily and seasonal variations in temperature may be great because of the small volume of water. Ice forms earlier and lasts longer in ponds than in lakes, freezing shallow ponds to the bottom in severe winters. Light penetrates to all depths, encouraging growth of vegetation except in high trubidity.

Plant Sere

The plant hydrosere, or pond sere typically contains the following stages and characteristic species:

Submerged vegetation: Water lily, pond lily, pondweed, smartweed, duckweed, and water hyacinth.

All except the last two are rooted in the mud, often at a depth of 2 to 3 meters and may have rhizomes from which long petioles extend to the leaves floating on the surface. Duckweeds and water hyacinths are unattached floaters, and cover the surface extensively in some localities.

Emergent vegetation (marsh): the dominant species are : cattail, reed, wild rice, and sawgrass. They invade water of over 1 meter depth, but in shallower waters or in secondary succession; they are replaced by the sedge meadow composed of sedge rush and spike rush.

Swamp shrubs: Buttonbush, alder, dogwood, swamp rose and sometimes shrubby willow and ctton-wood.

Swamp forest: Red and silver maples, elm ash swamp white oak, pin oak.

Suceeding stages depend on the climate region. In arid regions the swamp forest may be poorly developed, and grassland or desert vegetation may come quickly. In the mesic climate of the astern stages, and oak-history associes follows the swamp forest, replaced in turn by climax sugar maple-beech or mixed mesophitic forest. The hydrosere in the broad leaved evergreen climax area of south-eastern North American brings in cypress and a number of other unique species.

Animal Sere

Animals characteristic of marshes and ponds constitute a distinct pond-marsh biocies which extends into sluggish or base-leveled streams and the lilttoral zone of eutrophic lakes. Most animal species are restricted to single stage or community of the plant sere, but commonly occur in several stages in varying abundance and for activities. Fish for instance, feed in open water but spawn in shallow water among the emergent vegetation. Submerged, floating and emergent vegetation represent different levels or strain in a single biotic community and each stratum has about the same degree of distinctiveness as forest community strata. With the invasion of swamp shrubs and the ground level well above the water table most of the year, ponds and marsh species largely disappear, replaced by many characteristic new species. This animal community represents the swamp facies of the deciduous forest-edge biocies.

The swamp forest is often quite open at its outer margin, and forest-edge species remains common. But as this forest develops a closed canopy and drier ground stratum, it is invaded by the swamp facies of the deciduous forest biociation. In the ponds sere there is a succession of animal adaptations from aquatic to amphibious to, terrestrial.

Bogs

Bogs or moors typically develop in the hydroseres of cold northern regions; while marshes and swamps, which are markedly different from bogs, are characteristically southern in their location. Bogs commonly develop into a coniferous forest climax; swamps succeed to deciduous forest or other southern climax types.

Several thousand years ago, when glacial climates gripped the northern states, extensive bogs developed and have persisted as relic communities inspite of the warming of the climate. These bogs are slowly being replaced by pondmarsh species at equivalent seral stages by cliseral succession. Bogs occurring in the Great Lakes region are ordinarily small in area and have little or no drainage. There may be oxygen present in the open water of the larger bogs, but it is characteristically in very low concentration at the most in small bogs or in the marginal zones. Bog water has a distinct brown colour a low nitrogen content; a low temperature beneath the surface; a low pH at least in the marginal vegetated zones; and a low dissolved salt content. A false bottom is characteristic of bogs. It consists of finely divided plant material of light brown colour, held suspended in the water at varying depths below the surface. This false bottom may be extended down ward several meters before a true solid bottom is reached. The material disperses on slight disturbance and may render all the open water turbid. Ordinarilly, the surface waters are quiet and clear. Dead vegetation does not completely decompose; as it accumulates, it becomes compressed to form peat.

Plant Bog Sere

In the early stages of development of the bog, organic detritus may accumulate mostly in the deepest portions. As time goes on, however, a definite concentric-circle zonation of vegetation is established around the margin. As peat accumulates, each zone encroaches on the next inner; the in most shriks until all open water disappears. The area finally is covered with climax forest. In some bogs the first plant stage may be composed of floating vegetation (Nuphar, Nymphaea potamogeton, Sparaganium), but floating vegetation is often absent and the first stage is a sedge-mal composed of sedges cotton-grass, and buckbean.

The rhizomes of the sedges grow out into the water and become so interlaced that they form a floating mat. The next plant stage is dominated by low shrubs, which encroach on the floating mat. The leatherleaf, bog rosemary, laurels, labrador tea, sweet gale, and cranberries are important species. A high shrub stage commonly follows the low shrubs at such times as the mat becomes thicker or grounded. Common shrub species are holly, willow chokeberry, alders, and dwarf birch. The first tree of the bog forest to invade the shrubs is commonly the tamarack, but this species is now less common than formerly because of fire, logging, and depredations of the sawfly larvae Lygaeonematus erichsonii. Black spruce may either invade the shrubs directly or follow the tamarack. Later, the northern white cedar may become dominant and persist for a very long time, but the ultimate fat of the bog, upon addition of upland soil or the water table, is to be covered with the climax forest of the region.

In bogs that have a large body of open water, or an inflow of water entraining oxygen, and in which the pH is not extreme, invertebrate life comparable to that found is ponds and marshes occurs. True bogs, however, have little oxygen and a low pH, and many pond species do not appear. Mollusks are characteristically absent; sphaeriids may persist but their shells become very thin. Bottom organisms in general are poorly represented because of the tenuous physical nature of the substratum.

Desmids predominate among the phytoplankton, although dinoflagellates, Chlorophyceae, and Myxophyceae are common. Rotifers and a variety of Protozoa are the principal Zooplankters. Amphibians and reptiles are not characteristic of bogs, although the leopard frog is sometimes numerous on the sedge mat of bogs in Minnesota. Proceeding northwards boreal species of birds become progressively more prominent in the bog avifauna. The muskrat and beaver persist into northern Canada.

In general, the productivity and economic value of bog is very low compared with ponds and marshes. Liming experiments, calculated to improve productivity, are being made. Calcium combines with the humid colloids, which then flocculate and fall to the bottom. This clears the water, light penetrates deeper, pH is raised and greater algal, zooplankton, and fish growth is induced. Peat is a special product of bogs. It is of importance in northern Europe. It is cut out in blocks, dried, and used as fuel and for mixing in the soil of gardens and lawns.

Rock, Sand and Clay Sere

The origin of life was undoubtedly in the sea. Physiological adjustments were necessary before organisms could occupy fresh water. Although some organisms may have become air-breathers and ivaded land habitats directly from the sea, most evolution of terrestrial forms has doubtless come from fresh water. Relatively few major groups of animals have been successful in this invasion of land, the most notable being oligochaete worms; gastropod mollusks; many arthropods, especially the insects and spiders; reptiles and their evolutionary offspring, the birds and mammals

Adjustments to the Terrestrial Habitat

Living on land presents many problems. Our present concern is to analyze the ways in which animals have met these problems and to trace the succession of communities in the extreme terrestrial habitats of rock, sand, and clay. A variety of techniques have been developed for measuring environmental factors.

Gravity

In water, organisms counteract gravity by means of various floating and swimming mechanisms. Fluid buoyancy permits water-dwellers to attain huge size; consider the whale. A land animal on the other hand, must support its entire weight. Some terrestrial animals gain a modicum of support by burrowing into the soil. Others drag their bodies over the ground surface. But the supportive advantage they thus gain is costly in other directions, for they are slow moving and relatively helpless before predators. The animals best adapted to terrestrial life have evolved appendages in the form of legs or wings that not only raise the body above the ground but are also the means of more or less rapid locomotion and adroit movements over the surface or through the air. Terrestrial adaptation has involved the development of a tough body covering to hold fluids and internal organs in place; a skeletal frame work to give permanent shape to the body and, as a system of levers, to furnish means of locomotion; and powerful muscles to lift and move the heavy body.

Moisture

In sharp contrast to aquatic forms, terrestrial animals are not constantly enveloped with a continuous watery medium, with the limited exceptions of protozoans, nematodes, and other small organisms living in moist soil. Land animals lacking constant contact with the water medium, are faced with the problems of obtaining water and preventing excessive water losses from the body. Water becomes available to animals in varying amounts in the forms of rain, snow hail, frost, and fog. Whatever the form it arrives in, the significant things are the amount of free, liquid water added to the sustratum, accessible to plants and animals and the humidity of the air.

Considerable amounts of moisture are lost to organisms as runoff water flowing into streams, by evaporation back into the air, and as water bound in snow and ice. Evaporation of water from the earth's surface or from the bodies of organisms increases as temperature rises, air movement (wind) accelerates, and the amount of moisture already in the air decreases. When measured as grains per cubic foot or as millimeters mercury pressure, the actual amount of moisture vapour in the air is known as *absolute humidity*. This measurement is of less ecological importance than is *relative humidity* the ration of amount of water vapour actually in it to the quantity required to saturate the air at existing temperature and barometric pressure. Relative humidity is easily determind by means of sling or cog psychrometers, and may be continuously recorded with temperature by hygrothermographs.

The evaporation rate of water is more closely related to *saturation deficit*, than it is to relative humidity. Saturation deficits is a quantity which cannot be directly measured, it must be calculated. It is that additional amount of moisture required to saturate air under prevailing temperature, relative humidity, and barometric pressure conditions, commonly expressed as grains per cubic foot or as milimeters of mercury pressure. The most exact and desirble measurement affecting water evaporation is the *vapor-pressure gradient* obtaining between the organism and the surrounding air.

The gradient is positive if water molecules leave the organism at a rate faster than the rate at which the organism is absorbing them from the air, and negative if the reverse is true. The determination of gradient magnitude involves the measurements of body temperature of the organism and air, permeability of body membrances, and rate of air movement over the body surface. Water is obtained by a land animal by various devices. There may be some direct *absorption* through body surfaces such as occurs in the toad in moist soil and in some beetle larvae in moist air, this device is important in only a fes species.

Large mammals frequently travel several miles each day to water holes to imbibe drinking water. Many, but not all, birds require *drinking* water; some species, for instances quail and partridge, get it as morning dew on vegetation. Butterflies may frequently be observed drinking water from small pools. An important source of water is the *free water in food*, particularly in succulent vegetation and in the blood and body fluids of animals. Desert animals depend almost entirely for water on that contained in their food and on *metabolic water*, liberated when fats and carbohydrates, and to a lesser extent, proteins, are oxidized in their bodies.

Water is lost from the body through the skin and lungs as insensible moisture perspiration. Rapid and largely uncontrolled loss of moisture through the skin of amphibians, snails, annelids, and insect larvae is a limiting factor confining these animals to moist habitats or to activity only in times of high humidity. Adult insects and other arthropods, reptiles, birds, and mammals have evolved body surfaces of chitin and waxes, scales, or cornification of the surface layers of the skin that largely prevent uncontrolled loss of moisture. Moisture loss through the respiratory surfaces in these forms remains considerable however.

Water is also lost with the faces, although in some species much water is reabsorbed by the large intestine before the feces, are ejected. The amounts of water removed from the body by excretory organs particularly the kidneys varies directly with the amount of water intake and immersely with the amount lost through other devices. The kidneys are critical to maintenance of proper concentration of salts in the blood and body fluids. It is important that water intake balance water loss. Organisms are very seensitive to disturbances in body water balances and this factor is significant in determining the type of niche which a species comes to occupy.

In order that animals could exist in terrestrial habitats, they had to acquire the ability to carry on reeproductive activities in the absence of water. The chief reproductive adaptations involve the following: internal fertilization; a shell covering the egg to conserve moisture and salts; food provision to the embryo and young by yolk in the egg cell, placenta in the uterus of the mother or direct feedings by adults; reduction in number of young with more efficient parental care; reduction or elimination of free-swimming larval stages; and greater segregation of species into different niches to avoid interspecific disturbances.

Temperature

Aquatic animals are not ordinarily subject to temperature below freezing and are in a relatively stable temperature environment, but terrestrial species are exposed to highly variable temperature that may reach extremes of about —680 C and +550C. No single species is required to with stand such a wide range of temperatures, however. Optimum and tolerance limits vary from one species to another. No aquatic species has evolved control over its body temperature. Aquatic warm-blooded mammals and birds are derived from terrestrial forms.

A ability to maintain a constant body temperature has survival value in terrestrial habitats however, and consequently physiological mechanisms for homoiothermism developed independently in birds and mammals. All other land organisms are poikilotherms; that is, they have no physiological mechanism for maintaining a constant body temperature.

Some poikilotherms, such as bees have developed special behaviour patterns that enable them to mintain fairly constant conditions in the hive by cooperative efforts; some moths and other insects with hair-like body coverings maintain a degree of constancy in body temperature by shivering their wings; some lizards, snakes, and turtles are able to exert some control over their body temperatures by moving into and out of sunlit areas.

Rates of activity, food consumption, metabolism, growth and other physiological functions increase, to a certain limit with rise of body temperature. Homoiotherms maintain a continuous high rates of functioning because of their constant high body temperature, but the rate at which poikilotherms function varies with the temperature of their habitats.

Oxygen

Dry air 760 mm Hg pressure contains approximately 21 per cent oxyen, 0.03 per cent carbon dioxide, 78 percent nitrogen and traces of other gases. Oxygen is thus much more abundant, constant, and available at all times in air than it is in water. Oxygen availability seldom becomes a critical factor for land animals, with occasional exception of forms that live in the soil or invade high altitudes. Although terrestrial organisms have evolved simple moist chambers, branched trachael systems, or complicated lungs to replace the gills found in many aquatic forms the fundamental requirement of moist membranes for the exchange or oxygen and carbon dioxide between body fluids tissues, and surrounding medium remains the same.

The skin still serves this purpose in some terrrestrial forms annelids and some amphibians but in most forms the moist membranes are within the body. Internal placement decreases the loss of water through evaporation. The evolution and any ability to take oxyen directly out of the air apparently preceded the actual invasion of land, and may have been induced invasion in the pond and marsh habitat when oxygen dissolved in the water becomes reduced or absent during summer stagnant periods. The evolution of internal air-breathing organs was probably concurrent with the evolution of mechanisms to prevent excessive water loss from the exposed surfaces of the body.

Rock Sere

Stages in the plant sere on bare rock include lichens, mosses, herbs and grasses, shrubs, and forest. The species composition of each stage varies with the chemical nature of the rock, the prevailing climate, and the locality. In the first stage, various kinds of lichens compete for a foothold, but crustose types usually precede boliaceous types. Mosses such as foliaceous lichens as Cladonia follow foliaceous lichens; or may initiate the sere telescoping the earlier lichen stages.

Lichens and mosses soak up moisture in wet weather. They derive mineral nutrients from the underlying rock. Carbon dioxide secreted from the rhizoids forms a weak acid with water and dissolves the binding material of the small rock particles. Rhizoids may penetrate rock for several millimeters. These plants tap wind blown dust and obtain nitrogen from oganic compounds in it.

When the plants die, they become an addition to the accumulation of organic matter. Herbs, grasses, ferns invade to continue the crumbling of the rock and build up of the soil. Freezing and thawing of water may crack the rock, and in these cracks wind and water borne soil lodges and supports plants. Once shrub and tree roots get started in crevices, their growth exerts a powerful force, further splitting and crumbling the rock.

Animal Life

Animal life in the pioneer plant stages on rock is scanty. Ants and spiders roam over the bare rock, and insects of various sorts may stop there, temporarily. Spiders may construct webs and nests in rock crannies or among the lichens. Some tardigrades find preferred niches in lichens. Mosses offer a somewhat more substantial micro-habitat, but only those animals that can tolerate great extremes of flooding, dessication, heat ad cold can survive. Such forms are found in the rhizopod protozoans, nematodes bdeloidid rotifers, tardigrades, copepods, small insects and mites. They often have special means of attachment to keep them from being blown away by the wind, such as strong claws or cement glands on the feet, long bristle-like threads to entangle among the moss filaments; stickers or spines covering the eggs. Since wet periods are often too short to permit complete development, all stages must be tolerant of dessication, at which time activities and growth are largely suspended.

Animal life in general and land snails in particular are usually more abundant in vegetation established on soils delivered from sandstone, granite or volcanic rock. Calcium carbonate is a mineral essential to the metabolism of most animals and for building such skeletal structures as bones and shells.

Clay Sere

Erosion or calculated removal of overlying material in strip mining for coal may leave bare areas of clay exposed and impound small bodies of water, often with low pH values. These water bodies undergo a modified pond succession. In clay above pH 1.5 annual plants, of which smartweed is particularly important appear within a few weeks to 2 years; the higher the clay pH, the quicker the appearance of vegetation. Within 2 to 5 years there after sweet clover invades and develops nearly complete dominance over large areas. Sweet clover is a biennial and an exotic species unimportant in the sere in some parts of the country. Prior to its introduction, this stage in the sere on bare clay may have consisted of the perennial grasses still found in small scattered patches.

The first trees begin to invade early in the sere, our they are scattered and slow or growth, and do not attain dominance for 25 to 30 years. Willows occur in wet spots herb species of the first two plant stages disappear, for the most part, in the shrub stage. The Herb stratum now consists largely of wood nettle. Advance forest stages of oaks, hickories, basswood and sugar maple will likely invade in the future, as they occur now in adjacent areas.

Animal Life

The number of invertebrate species tends to increase as the sere advances, although not always regularly. In a study of a formerly stripmined area, 18 species were found to be important in the annual stage, 41 species in sweet clover, 40 species in shrubs, 32 species in the early forest stage, and 67 species in the unplan climax. More species would be found in advances stages because of the greater variety of niches available. Thus, in the initial bare area there is only the ground stratum, in the annuals and sweet tree strata.

Since the early forest is on a flood plain, the ground is frequently swept by floods, and the shrub stratum is poorly represented. The climax forest has all strata, richly developed, and possesses the greatest number of animal species. There is an increase in the abundance of individuals per m^2 with the progression of the sere: annuals 268; clover, 531; shrubs, 532; early forest, 748; climax, 2445. Beetles, spiders, ants and mites are the most abundant animals in the annuals stage, and along with aphids remain most abundant also in the sweet clover community.

Grasshoppers are fewer in number but especially characteristics of the first three stages; they practically disappear in the forest. Earthworms are absent in the annuals and scarce in the sweet clover, as are the springtails; as the amount of soil humus increases with the development of the sere, both groups become more and more numerous. Snails first appear in the sweet clover stage and increase in importance in the forest stages. Starting with bare ground and continuing through early and late shrub stages to bottom land forest, there is a progressive increase in bird species diversity/density and biomass. The first small mammal to invade the area is the prairi deer mouse followed by cottontail rabbit and fox-squirrel.

Floodplain Sere

A stream continuously deepens its channel, thus lowering the water table of the surrounding land. At times of heavy rain, the channel may not be deep enough to carry the discharge and the river overflows its banks. In mesic climate this commonly happens three or four times per year, usually during late winter and spring, and each flood may last up to several days.

The more extensive floods, persisting for long periods, come at less frequent intervals. The low land area between the river and the bluffs on each side is called the *flood plain*. In the course of time, the river in its erosion meanders back and forth across the floodplain, cutting new channels and the surrounding upland. The higher the terrace or the farther area is from the channels and abandoning old ones, and frequently leaves a sequence of terraces between its present channel the less frequently and for shorter periods is it flooded, so that there is commonly a sequence of plant communities that constitute the plant sere.

Plant communities

Gravel, sand, or silt is deposited on the inner side or river bends. Attached acquatic vegetation may occur in the water. On land, such herbs as smartweed, cocklebur, ragweed, beggar's ticks occur. At some bends small sand dunes may occur, displaying their characteristic plants and animals; usually this stage is narrow at most, and may be entirely absent. On sandly islands in the river or on sandly shores the sandbar willow often forms dense, shrubby thickets. The first tree stage is frequently black willow mixed with eastern cotton wood, and sometimes silver maple. On the floodplain of the Canadian River in Oklahoma, the sere proceeds next to an edaphic subclimax of either tall grass prairie or elmoak. The climax on the surrounding upland is mixed prairie. The normal sequence of stages in this region has become considerably modified by the extensive ecesis of the exotic tamarisk tree, introduced from Asia.

On the Mississipi floodplain in western Tennessee the mature cottonwood-willow associes contains an abundance of wines of several species that from such tangled masses as to be almost impenetrable. The next stage is one in which sugarberry sweetgum, American elm, and American elm, and American sycamore predominate; several other species are present in small numbers. This leads to an oak-hickory stage that includes a complex variety of species, and eventually to the regional climax of western mesophytic forest.

Cypress becomes part of the composition of the early floodplain forest around the edge of small oxbow ponds or other standing water. The schedule for this sere, the time from the start to the beginnings of dominance by each successive plant community has been estimated as follows: sandbar willow, 3 years; cottonwood-willow, 35 years; sugarberry-seetgum, 82 yeaars; early species of oaks and hickories, 260 years; intermediate species of oaks and hickories, 350 years; eraly climax of oaks and tuliptree, 440 years; full development of the climax, 600 years.

Elsewhere in the eastern United States, the cottonwood-willow stage gives way to a narrow zone of a sycamore. Two or three species of elm, white ash, and boxeleder follow; then a mixed forest that includes black walnut but ternut, blackmaple ohio buckeye red mulberry, American basswood tuliptree, and hackberry; next an oak-hickory stage; and finally the beech-sugar maple climax. The herb and shrub strata are usually well developed in mature floodplain forests. Telescoping or skipping of stages is not uncommon in this sere, since variation in ground level or in height of terraces is considerable and the transition between heights is often abrupt. The later stages occur only on the very oldest terraces and may be hard to find at all.

Animal Life

In the bare areas, in the herbs, and among the invading trees occur such beetles as *Heterocerus pallidus* and *Bembidion laevigatum* that feed on the algae and detritus present on the shore. They make their burrows in sand. Fly larvae, a cocklebur weevil, a cocklebur mirid, and a cocklebur fly also occur. The tiger beetles *Cicindela hirticollis, C. cuprascens,* prey on the ground species and may even dig them out of their burrows. Spiders, ground beetles, and rove beetles invade from higher stages.

In the herb stratum and in the shrubby growth of willows, adult midge flies and other flies are sometimes very abundant. Tarnished plant bugs, 12-spotted cucumber beetles, and other insects of open-area habitats are present, and there in invasion of various species from the forest itself. The animal life of the floodplain forest in much the same as that of the deciduous forest in general and does not need to be discussed here except for its unique features.

Annelid worms make their appearance in the ruderal stage, become very abundant in the moist soils of the elm-ash and mixed floodplain forests, then decrease in numbers in the direr soils of the late seral stages. The occur mostly in the first 5 or 10 cm below the surface in moist soil, but up to 30 cm or more in dry soil. During the winter they keep below the frostline, and in very dry weather they roll up in small knots. Ten species occur in the floodplain of the Snagamon River in central Illinois, and each species has its particular range of moisture requirements between the river's edge and the upland forest. Snails and slugs and moisture-living animals and occur in large animals and great variety in floodplain forests; it is not hard to find 15 to 20 species with a little searching. *Mesodon thyroidus* in a common snail, and on a floodplain in central Illinois an average population of 6.3 individuals per m^2 was found during the autumn. This amounts to a biomass of living flesh (shell ex-cluded) of 15.8g/m^2 (141 lb/acre). Snail flesh is an important source of food for such small mammals as the short tailed shrew.

The word ecology, derived from the Greek word: oikos meaning habitation, and logos meaning discourse or study, implies a study of the habitations of organisms. Ecology was first described as a separate field knowledge in 1866 by the German zoologist, Ernst Haeckel, who invented the word oeckologie for "the relation of the animal to its organic as well as its inorganic environment, particularly its friendly or hostile relations to those animals or plants with which it comes in contact." Ecology has been variously defined by other investigators, as "scientific natural history", "the study of biotic communities", or "the science of community population"; probably the most often given: a study of animal and plants in their relationship to each other and to their environment.

Ecology is a distinct science because it is a body of knowledge not similarly organized in any other division of biology; because it uses a special set of techniques and procedures; and because it has a unique point of view. The essence of this science is a comprehensive understanding of the import of these phenomena:

1. The local and geographic distribution and abundance of organisms (habitat, niche, community, biogeography).
2. Temporal changes in the occurrence, abundance, and activities of organisms (seasonal, annual, successional, geological).
3. The interrelations hips between organisms in populations and communities (population ecology).
4. The structural adaptations and functional adjustments of organisms to their physical environment (physiological ecology).
5. The behaviour of organisms under natural conditions (ethology).
6. The evolutionary development of all these interrelations (evolutionary ecology).
7. The biological productivity of nature and how this may best serve mankind (ecosystem ecology).
8. The development of mathematical models to relate interaction of parameters and predict effects (systems analysis).

Ecology is one of the main divisions of biology, the other two being morphology and physiology. The emphasis in morphology is on how organisms are made, in physiology on how they function, and in ecology, on how they live. These divisions overlap broadly. To appreciate fully the structure of an organ, one needs to know how it functions, and the way it functions is clearly related to environmental conditions.

The morphologist is concerned with problems of anatomy, histology, cytology, embryology, evolution and genetics; the physiologist, with interpreting functions in terms of chemistry, physics and mathematics; and the ecologist, with distribution, behaviour, populations and communities in relation to the environment (ecosystems).

The evolution of adaptation and of species is of mutual interest to the ecologist and to the geneticist; biometeorology is a connecting link between ecology and physiology; and system analysis inter-relates ecology and mathematics. All areas, in the final analysis, are simply different approaches to an understanding of the meaning of life.

Subdivisions

Ecology may be studied with particular reference to animals or to plants, hence animal ecology and plant ecology. Animal ecology, however, cannot be adequately understood except against a considerable background of plant ecology. When animals and plants are given equal emphasis, the term bioecology is often used. Courses in plant ecology usually dismiss animals as but one of many factors in the environment. Synecology is the study of communities, and autecology, the study of a species.

There is some confusion in these terms since Europeans commonly use "ecology" in narrower sense—meaning the environmental relations of organisms or of communities. The broader study of communities, including species interrelations and community structure and function as well as environmental relations (synecology), is generally termed "biocenology" or "biosociology" by Europeans. We shall survey the fundamentals and basic facts of ecology as they relate to animals and have application to man. We will study community ecology, the local distribution of animals in various habitats, the recognition and composition of community units, and succession; ecosystem dynamics, the processes of soil formation, nutrient cycling, energy flow and productivity; population ecology, the manner of population growth structure, and regulation; evolutionary ecology, the problems of niche segregation and speciation; and geographic ecology, concerned with distribution, paleoecology, and biomes. We will be interested throughout the text with how organisms respond and adjust physiologically to the physical factors of their environment, but a full study of physiological ecology must be left to another time and place. We will also be concerned throughout with systems ecology that is the possibility of translating ecological concepts into mathematical models, although we will not go deeply into the actual statistical manipulations.

History

That certain species of plants and animals ordinarily occur together and are characteristic of certain habitats has doubtedless been common knowledge since intelligent man first evolved. This knowledge was essential to him for procuring food, avoiding enemies, and finding shelter. However, it was not until the fourth century B.C. that Theophrastus, a friend and associate of Aristotle, first described interrelations between organisms and between organisms and their environment. He has, therefore been called the first ecologist.

The modern concept that plants and animals occur in closely integrated communities began with the studies of August Grisebach, a German botanist, in 1938; K.Mobius, a German investigator of oyster banks, in 1877; Stephen A.Forbes, an American, who described the lake community as a microcosm in 1987; and J.E.B.Warming, a Danish botanist, who emphasized the unity of plant communities in 1895 C.C. Adams recognized and described many animal communities in his ecological surveys of northern Michigan and of Isle Royale in Lake Superior, published in 1906 and 1909. V.E. Shelford presented a classic study of animal communities in temperate America in 1913, and Charles Elton published an outstanding analysis of community dynamics in 1927. Although an appreciation that the whole community is one biotic unit, rather than one unit plants and another of animals, may be described in the writings of some early investigators, the fact has been brought to modern emphasis in the work of F.E. Clements and V.E.Shelford, especially in their Bio-ecology, published in 1939.

Much interest has been stimulated in recent years by D.Ramon Margalef, Robert Mac Arthur, and others for analyzing the structure of these biotic communities, particularly in respect to such phenomena as species diversity, niches, and how they come about through evolution. Geographic ecology, in the modern sense, dates from the generalizations on the worldwide distribution of animals made by the French naturalist, Georges L.L. Buffon, and the explorations of the German naturalist Alexander von Humboldt. There was lively interest and many important contributions in this general field during the nineteenth century; notably, the life-zone concept of C.Hart Merrian needs special mention. During the present century the concept of biotic provinces is identified with L.R. Dice and the biome concept with F.E. Clements and V.E. Shelford.

The broad survey of ecological animal geography made by R. Hesse in 1924 exerted considerable influence, and this treatise was later translated into English and revised by W.C. Allee and Karl P. Schmidt. The study of population dynamics, so important in modern ecology, dates back at least to Malthus, who pointed out in 1798 the limitation to population growth exerted by available food. Darwin, in 1859, recognized the importance of competition and predation in developing his theory of evolution. Pearl, 1925, analyzed mathematically the characteristics of population growth, and Lotka, 1925, and Volterra, 1931, developed theoretical mathematical equations to show the manner in which populations of different species interact. These studies led to the classic experiments of Gause, 1935, with interacting populations of predators and prey. Nicholson's publication in 1933 stimulated much thinking concerning the factors that stabilize populations at particular levels.

Andrewartha and Birch, 1954, emphasized the importance of climate and other factors on determining the size of populations. Physiological ecology had its historical beginning in the correlation of biological phenomena with various temperature stimulated by Galileo's invention of a hermetically sealed thermometer about 1612 A.D. The French naturalist Reaumur summed the mean daily temperatures for April, May and June in 1734 and again in 1735, and correlated the earlier maturing of fruit and grain during first year with the greater accumulation of heat.

A discovery of parallel significance was of oxygen in 1977 that it was an essential part of air, Claude Bernard, another French Physiologist, enunciated the principle of homeostasis in 1876. This concept originally referred to regulatory mechanisms which maintained the "internal environment" of the body constant in the face of changing external conditions. Later, the concept came to be applied also to maintenance of community interrelations. The development of animal behaviour or ethology may be traced back through the natural history of ancient times. More recently the 13 volumes of Thierleben, prepared by A.E. Brehm during the period 1911-18, are noteworthy, H.S. Jennings, 1906, and Jacques Loeb, 1918, made valuable contributions to the understanding of the behaviour of invertebrates.

Precise modern techniques and concepts as applied to vertebrates began to take form about 1920 with the development of banding or marking of individual animals by S. Prentiss Baldwin and the recognition of territories in the nesting of birds by H.E. Howard. The formulation of the concept of releasers as controlling instinctive behaviour by Wallace Craig, K. Lorenz, and E. Thinbergen has produced a profound effect on present-day thinking. Ecology, then, is of comparatively recent development as a distinct science, but its roots extend well back into the past.

An early comprehensive treatment of the subject is Principles of Animal Ecology by Allee, Emerson, Park and Schmidt published in 1949. Since ecology is a young science, it should be emphasized that its concepts and techniques have not become standardized and that there is opportunity and stimulus here for many new investigators.

Ecological Concepts

Ecology deals with 'the web of life' that entangles: (a) every species with the lives of others; and (b) each species with its non-living environment as a whole and each element or factor of that environment. The term organisms refers to all that is living and the term environment denotes everything external to the organisms and includes the physical environment such as climate and the biotic environment. Odum has preferred to define ecology as the 'study of the structure and function of nature'.

The total physical environment of any living organism is the *biosphere*, and parts of which are of direct or immediate importance to the organism constitute the 'effective environment'. The complex constituting the organism and the physical environment has now come to be known as *ecosystem*. Ecology then, is the study of the inter-relationships within ecosystems. The environment devoted crop plants is the *agroecosystem*, where both animals and plants in the area are directly or indirectly influenced by each other.

Ecosystems are not closed systems because there is always an interflow of matter and energy between adjacent ecosystems. The principle of the ecosystem has also brought forth very clearly that the 'whole' is more than the 'sum of its parts'. The interdependence of the organism and the environment must provide the minimum requirements and nothing that is lethal. At the same time living organisms should also be able to successfully adapt themselves to their surroundings. It must be realised that the natural pressures of the environment continue to act on the living organisms and ecological studies relate to the ways by which they—singly or collectively—respond to these pressures or changes.

The inability of many animals to cope up with the surroundings, has naturally led to their extinction from the surface of the earth. It may be said that dispersal or the ability of organisms to spread from their place of origin to another area, is a basic aspect of population ecology and without proper means of dispersal, many animal populations would have been eliminated over a period of time due to modifications of the environment. In environmental studies it is usual to refer to the *microenvironment* and the *macroenvironment*. Maelzer (1965) suggests that environment should be defined as the sum total of everything that directly influences the animals' chances to survive and reproduce.

While the microenvironment is the intimately local and immediate surrounding of the organism, the microenvironment is the sum total of the physical and the biotic conditions existing external to the organism and its microenvironment. Both contain many factors affecting the establishment, growth and reproduction of animals and plants. Some of these factors relate to oxygen and carbon dioxide pressures, barometric pressure, wind movement, solar radiation, etc. The two divisions of the environment may differ drastically or only slightly from each other with regard to these factors; for instance, the temperature inside the burrow of a rat is different from that outside it; or within a forest the temperature beneath a fallen log or a stone would be nearly half compared with that of the air, while the temperature at their surface would be much higher.

The restricted nature of such microhabitats naturally calls for special adaptations and the term *microclimate* is often used for the physical conditions of microhabitats. To the ecologist the species is his fundamental unit and the environment generally harbours more than one animal or plant species or species populations. [Ecological studies may deal with the study of individual organism or species in relation to their habitat (*autecology*) or the study of associations of organisms in relation to a particular area or biotope of habitat (*synecology*).

While synecology is the ecology of the biotic community, demecology refers to the ecology of populations. In autecological studies the assessment of the intrinsic capacities of organisms which enable them to cope with environmental conditions is important. The term *ecological valency* has been used to refer to the adaptive range of both the individual and the species. (An assemblage of populations inhabiting an area possessing some uniformity in physical feature and begetation constitutes the *community*.) (While a population is formed of a number of animals and plants inhabiting an area, a community is an aggregate of several species of plant and animal population—the *biotic community* which forms a distinct ecological unit.)

Communities differ according to the environment and the nature of the community is an index of the environmental conditions. The place that an animal occupies in a biotic community which express its functional status, is called *'niche'*. The niche may be better defined as the place that an animal occupies in a biotic community in relation to its food and enemies. It results from the animal's peculiar structural adaptations, physiological adjustments and special behaviour patterns. Every animal exhibits a specificity with regard to its niche and no two animals can occupy the same niche permanently for a long time in the same locality.

Appropriate analysis would reveal an interconnection of the different niches into a network which holds the different animals together as a community. This brings about a stratification of the community on the basis of their mode of food capture. It can easily be understood from a study of the relationship of organisms of a community to the food web in a particular environment. *Trophic level* is the term applied to such relationships on the basis of food, and through successive trophic levels the primary stock of energy as obtained by green plants is continually circulated. The chain or sequence of events in which the herbivores consume plants, carnivores feed on the herbivores (and saprophytes feed on all organisms) constitute the *food chain*.

All biotic communities are in a state of constant change resulting from the actions and the coactions of the organisms or of the external environment. Changes occur in a more or less orderly way so that in course of time the community is replaced by another community. In general, it may be said that communities after they come into being, grow, mature, become senescent and as they die their area become inhabited by other communities. This progressive sequence of replacement of communities in a given area continues until a more or less stable community is established. This has been termed *ecological succession*.

Modern ecology relies on a variety of disciplines like physics, chemistry, mathematics, meteorology, climatology, geology, geography, etc. An essential aspect of these studies is concerned with energy transformations occurring within ecosystems. The quantity of incident energy from the sun per unit area of the ecosystem and the efficiency of conversion of this energy by other organisms is a basic principle. For the effective functioning of ecological systems energy transferring mechanisms from organism to organism as observed in food chains, are of prime importance. *Ecological energetics* deals with this aspect of ecological studies.

The autotrophs (plants) and the heterotrophs (animals) utilise this energy. Such interconversions of energy closely follow the laws of thermodynamics and it may be said that the solar energy entering the ecosystem is equivalent to the heat energy leaving the system. Studies such as ecological energetics provide avenues for an inter-disciplinary approach to ecological studies. Significant contributions of other disciplines towards a better understanding of ecological principles have given rise to ecological genetics, palaeoecology, ecophysiology, agroecology, chemical ecology, ecogeography, and so on. An ecologist recognises a kind of genetic plasticity in the case of every animal. In any environment, only those animals that are favoured by the environment survive. (The branch of ecology dealing with genetics in relation to ecology is ecological genetics.

Palaeoecology deals with the movements of biotic elements based on palaeoentological evidence, which provides information about ancestral organisms and environmental conditions existent in the past.) While it is difficult to draw a line between ecology and physiology, the factors of the environment have a direct bearing on the functional aspects of organisms. All the same it may be said that ecology is concerned with supraindividual units—communities and populations—while a physiological approach generally deals with the individual animal or plant. As such (ecological studies having a direct bearing on the survival of populations as a result of functional adjustments of organisms, come under ecophysiology.

Similarly the adaptations of animals or preference of particular organisms like insects to particular chemical substances, come under chemical ecology. Of considerable interest to terrestrial ecologists is the study of soils or pedology, in particular their acidity, alkalinity, humus content, etc., and their influence on the organisms). The ecogeographic approach emphasises the role of the environment in animal distribution. It is well known that zoogeography relages to the structural and functional relations of animals in space, which form the immediate environment of the individuals as well as the populations.

The importance of faunal analysis or faunal complexes is recognised by some zoogeographers, and Uvarov considered *ecofaunas* as the lowest units of which a geographic fauna is made up of. Another aspect of ecogeography is seen in the dispersal or spacing of fauna as a result of multiple interactions between the individual and the environment. While the environment offers considerable resistance, the individuals tend to adapt themselves, exemplifying the role of selection in the evolutionary process. Dispersal within the area of distribution pertains to the concept of population ecology, while dispersal outside limits of normal range of the species is essentially a zoogeographic concept.

With the fast rate of multiplication of human population, problems of food and space will naturally arise. Man has been known to control his environment successfully to meet his needs, but indiscriminate control of insects and other animal pests through use of insecticides, the release of massive quantities of radioactive debris, discarded chemicals and industrial waste into rivers result in atmospheric and aquatic pollution, which may have short and long term ecological effects.

General Ecological Considerations

From very early times man has advanced his own ends by the modification of natural communities and ecosystems, sometimes in minor ways to enhance the yield of a desired commodity, but with the rise of agriculture, civilization and modern technology in ever more drastic ways. Such modifications are most likely to be productive and stable when environmental limitations of the natural ecosystem are well understood.

Ecosystems and their functioning

Any assemblage of plants and animals able to exist within an area will in time form a biotic community. In this community the different species tend to interact with one another and to modify the conditions of life within which each exists. They therefore develop the inter-relationships and inter-dependences which constitutes an ecosystem. In a forest grove or woodland, for example, the larger trees—the dominants—influence the environment in which all other species live.

Plants which cannot grow in their shade or tolerate the competition for soil, water and minerals provided by their root systems, or the various chemical substances that they may release into the soil, are unable to find a place within the community dominated by these large trees, or will be confined to places where such trees cannot maintain themselves. Thus, the trees exert an obvious influence on other plants. However, smaller plants may modify the environment in various ways to enhance or detract from its capacity to support the larger trees. The inability of various pine trees to thrive in the absence of the mycorrhizal fungus that attaches itself to their roots has long been known. Similarly, legumes' roots provides a home for bacteria which in turn manufacture the nitrates on which legumes depend.

The animal components of such a biotic community also modify the conditions of life for the plant (and are, in turn, affected by the plant community). The interrelationship (symbiosis) between the Yucca plant and the yucca moth is a well studied example; the plant supplies food for the moth, but is dependent on the moth for fertilization and perpetuation. Similarly, large grazing animals affect the populations of the plants on which they feed, and are dependent upon the quality and quantity of plant growth for survival.

Carnivores may also limit or otherwise modify the populations and the behaviour of herbivores, and thus have a secondary impact on the kind and intensity of grazing. It is convenient to draw a distinction here between 'natural' and 'artificial' communities. The former consist of wild, naturally occurring species of plants and animals able to maintain themselves in the absence of man. Artificial communities are characterized by species introduced by man or favoured by human modification of the environment, and are unable to exist without continued human assistance or interference.

Neither the word 'natural' nor 'artificial' is entirely satisfactory, since man is part of nature, and all communities, whether strongly influenced by man or not, are also a part of nature. Furthermore, it is difficult to find any area of the earth in which some human modification of the environment has not occurred, if no more than a minor changes caused by an increased frequency of certain radio isotopes, or chemical fall-out from air pollution.

Both categories, though useful to distinguish for some purposes, always exist as part of an *ecosystem*, which is essentially a biotic community in interaction with its physical environment. The physical environment or the abiotic component is composed of sunlight, atmosphere, water and soil or rock. All of these physical elements affect the living organisms, and all are in turn affected by them. Plants, for example, obtain their energy from sunlight, their basic food from air and water, supplemented by chemicals from the soil or rocks.

Plants, however, differentially reflect, refract or absorb various wavelengths of sunlight, and thus modify it; by using carbon dioxide and by giving off oxygen and water vapour they modify the atmosphere; by removing various chemicals and by adding others they modify the soil. All plants and animals always exist as part of an ecosystem. Man, like other animals, is dependent upon the ecosystems in which he exists despite his high mobility which enables him to move from one to another (or to extract products from one ecosystem and export them to another), and despite the technology which permits him to create major modifications in any ecosystem.

All ecosystems must have certain component parts or functions. Thus, they must have a source of energy (usually sunlight) Biotic Component have the following organisms:—

i) *Producers*:— They convert the sun's energy into chemical or food energy (usually green plants).

ii) *Consumers*:— They depend upon the producers for food. They may be herbivores, carnivores etc.

iii) *Decomposers*:— They take the chemicals built up by the green plants and break them into simpler forms in which they can be re-used. These include the so-called reducers organisms, such as bacteria and fungi, that act in the breakdown and decay of dead plants and animals.

From the functional point of view ecosystems has two components:

a) *Autotrophs.* (Self nourishing)

The organisms can fix light energy and can use simple inorganic substances and build complex organic substances.

b) *Heterotrophs* (Other-nourishing)

In these organisms there is utilization, rearrangement and decomposition of complex substances into simples ones.

The substances of the ecosystem depends upon the diversity of species in the ecosystem and keeping in balance the minerals between the biotic and abiotic components of the ecosystem. An ecosystem is self maintaining system except that it depends on the sun as external source of energy.

The living components of ecosystems are the plant and animal species (including micro-organisms) that make up the biotic community. Each species is adapted to a particular role in the ecosystem known as its *ecological niche*. Each depends for its existence upon the presence of a suitable habitat, comprising other species as well as necessary components of the physical environment. The functioning of the ecosystem is dependent upon the presence of a suitable combination of species each of which performs a specialized task within the total system.

In a natural ecosystem sun energy will be captured by trees, grasses, or other herbs, and converted into proteins, carboy drates and other food components. These will be consumed by plant-eating mammals, birds, insects and a great variety of other animals, which, in turn, will provide food for meat-eating animals. All ultimately provide food for the organisms of decay which will break down plant and animal materials and restore their chemical components to the soil. Energy will pass through an ecosystem in a one-way path. Of the solar energy reaching the earth, only a portion will be stored by plants; often less than 1 per cent of the total sunlight energy falling on a vegetated area is retained as chemical energy by the green plants. Of the total energy thus available in plant tissues for consumption by animals, usually less than 20 per cent will be stored as chemical energy in an animal's body tissues.

Eventually, after supporting life for a time, most of the energy is lost to the ecosystem. Some however; is stored in log-lived organic materials such as tree trunks, or in dead organic materials preserved from decomposition—those that eventually form organic deposits of one kind or another, including peat and, over longer periods of time, oil, coal and natural gas. By contrast, chemical nutrients flow through and ecosystem in circular pathways, from soil to plants and animals and back to soil, being reused or recycled again and again.

The Biosphere and its Functioning

All of the ecosystems of the earth, together, form the biosphere. The biosphere is that portion of the earth within which life exists. It includes all oceans and freshwater, the lower layers of the atmosphere and the outer skin of the earth's crus—the rocks and soil of the earth's surface. Within the biosphere all of the functions and structure described for any one ecosystem exist. Green plants capture sun energy and combine it with chemical raw materials from soil, water and air.

The food they produce supports all animal life, including the decay organisms which return it to the soil for plant use once more. Man is a part of the biosphere and depends on its continued functioning for his own existence. He can modify or even destroy any one ecosystem. He cannot, however, risk major modifications of the biosphere except at the risk of his own extermination. Thus, the continued production of plant materials, whether wild or cultivated, is the basis for the nutritional support of man as well as all other animals. The continued functioning of green plants is the source of atmospheric oxygen on which man and other animals depend.

The continued functioning of the reducer organisms is the means by which the chemicals in human wastes or in the bodies of plants and animals are made available for further use by living things. A breakdown in any of these biospheric systems would imperil human survival. Man has been able to take major risks in modifying local ecosystems. He cannot afford major risks in dealing with the functioning of the biosphere.

Factors Influencing Population Growth

The growth of any species population, whether it be of trees, wild animals or men, depends upon the excess of births (natality) over deaths (mortality). For any local area or ecosystem this may be influenced by immigration, movement into, or immigration, movement out of, the area or system concerned. For the biosphere as a whole, however, there is no immigration or immigraiton.

Populations, unless otherwise checked, tend to grow rapidly when there is an abundance of space and of the materials that the species requires for its subsistence. They grow much more slowly when space or essential materials are in short supply. They level off or decline when space or materials become limiting. A population growing at the maximum rate for which that species is capable is said to be growing at its biotic potential rate.

Such a rate of growth is only exhibited when conditions are most favourable for the species—when natality is at a maximum and mortality at a minimum operating to prevent a species from maintaining a biotic potential rate of increase is the environmental resistance. This is the sum total of all of those factors that cause mortality or decrease natality.

When biotic potential equals environmental resistance, the species population will not grow. Most animal populations have long been established in the area which they occupy. They seldom show either a continued growth or decline unless the conditions of their environment change drastically. Most show a fluctuation around some mean level which is determined by the ability of their environment to support their members.

Population of animals (or of men) introduced into an area previously uninhabited by the species concerned but otherwise suitable for it, tend to show an increase in a manner which resembles that of an S-shaped or logistic curve. This is at first marked by a rapid (exponential) rate of increase, and then by a period of levelling off towards some relative stability. Previously established populations, when freed from some factor of environmental resistance, may show a similar growth pattern. There are many modifications of the shape of the curve, depending on the species and the conditions of the environment, but the eventual period of levelling off, through increase environmental resistance, is characteristic of all growth curves.

Limits to Population Growth

All populations must sooner or later level off their curve of growth. This levelling-off can result from the behaviour of an animal species. Some species are intolerant of crowding and produce relatively fewer young as their habitat becomes more fully stocked. Other species, however, are more directly limited by the action of factors outside of the population which cause increased mortality. Thus population growth must inevitably be limited by the environment. No environment is without limits and the biosphere is limited both in extent and resources.

If there were no other restrictive factors, ultimate limits to the increase of any species would be imposed by the amount of sunlight energy reaching the earth. In fact, however, other factors are always limiting. For plants, limits to growth would be set primarily be the extent of their ability to convert sunlight into food energy, or by the availability of carbon dioxide in the atmosphere. In the operation of natural ecosystems, however, either water, or chemical nutrients present in soil or water, run out before the limits of carbon dioxide or photosynthetic efficiency are reached.

Animal populations must ultimately be limited by the availability of the food that eat, which depends on the ability of plants to continue supporting herbivorous animals. In practice, most animal populations tend to be limited by other factors before the absolute limits of food supply are reached. These other factors are most notably behavioural, those responses of animals to the presence of other members of their species that prevent an excessive accumulation of individuals in any one area. Interspecific predation can also be a limiting factor of some importance.

Human populations have attained comparative freedom form the limitations imposed by any one ecosystem—the exceptions being the remaining primitive groups of people. However, human populations cannot escape the limitations imposed by the resources of the biosphere. Of particular relevance is man's efficiency in converting those resources for use and in recycling waste products or by-products for ultimate reuse.

Carrying Capacity

The environmental limits to the growth of any one species population determines the carrying capacity for that species. Carrying capacity is a measure of the number of individuals of any species that a particular environment can support. Carrying capacity can be considered as having several levels:

1. an absolute or maximum carrying capacity which is the maximum number of individuals that can be supported by the resources of the environment at a subsistence level (this level can be termed the *subsistence density* for that species);
2. the level at which a species population is normally held by the influence of other species living in the same environment—those that hunt or prey upon the species, and those that cause disease or parasitic infestation (this level has been termed a *security density*, or threshold of security, because populations below this threshold are relatively secure from predation or disease; this is necessarily less than the subsistence density);
3. a level, which is generally considered more desirable by those concerned with the health or productivity of the species involved, termed an *optimum density*. At this level individuals in the population will have available an adequate supply of all essentials for existence and, in consequence, will show abundant individual growth and health not limited by shortages of any essential requirements. Such an optimum density can only be maintained by strong limitations on growth imposed by the behaviour of the species concerned (self-limitation) or by removal of individuals in excess of this density by the action

of other species through predation. In the latter case optimum density and security density will be the same. These are not necessarily fixed levels, but will fluctuate upwards or downwards.

Intelligent management of domestic animals for maximum production of meat, milk, etc., involves holding their populations at or near an *optimum density*. At this level, their productivity will be at a maximum and the highest yield can be obtained. Under primitive conditions in Africa and Asia, for example, domestic herds were often held at or near a *security density* by the action of predators (human or wild animal), or by diseases and parasites.

Under modern conditions, with predation and disease minimized, and where they are not properly managed, domestic herds often increase to a *subsistence density*. Under these conditions malnutrition is common, disease is more prevalent and productivity is low. Obviously, man has similar choices with his own populations. In primitive times the operation of predation, disease, etc., often held local populations at or near a security density.

With the removal of predators, and the control of disease, populations can either increase to a subsistence density, a level where human suffering and general misery will be maximum, or they can be levelled off by self-imposed restrictions on growth, at or near an optimum density. This would allow maximum individual opportunity for health, happiness, and choice of high quality life styles and environments. These choices would normally be available to all human populations at some period in their growth. Now for the first time, however, the whole humans species faces possible population increase to levels in excess of the capacity of the biosphere to maintain them, the biosphere itself being strained by rising demands for material resources and products. Carrying capacity is not fixed. It fluctuates naturally with weather and climate, and the operation of other natural facros such as fire, floods, earthquakes and vulcanism. In any one area it can be increased to some maximum level which is determined by the rate at which that environmental requirement which is in shortest supply can be provided. Such a requirement is known as a limiting factor. A limiting factor is that substance or quality in the environment, the supply of which is least abundant in relation to the needs of the animal or plant concerned.

Limiting factors and their operation

Limiting factors are often classified in two categories: density dependent and density independent. The effect of the former on a population increases as the population increases in density. It is essentially a stabilizing influence on growth and causes a population to level off at carrying capacity.

A density-independent factor, conversely, affects many or a few individuals without reference to the population level. Food supply is generally density dependent—the more there are to eat it, the less there is for each individual, and the grater the effect of food scarcity. By contrast, a flood is density independent since it may wipe out an entire population of a species, whether there are few or many. The more important limiting factors as follows.

(a) Climatic and Atmospheric Factors

These operate in a variety of ways to affect species populations. To begin with, any species has limits of tolerance and some optimum range of tolerance for such factors as sunlight, temperature, humidity, rainfall or wind. If a local climate normally exceeds the limits of tolerance for a species, then the species will not occur in that area; if it exceeds the optimum limits, the species will not thrive there. If the local climate only rarely exceeds the limits of tolerance, then a species may temporarily occupy that area, but will be eliminated in those years when the climate becomes extreme. Assuming that the climate is favourable to the establishment of a species, then a number of climatic factors may influence population growth.

Changes in temperature—years that are warmer or colder than normal—may permit a species to thrive and increase, or they may cause a decrease and permit its survival only in the most favourable sites. Changes in rainfall and humidity, dry years or cycles and wet years or cycles, have major effects. Fluctuations in temperature and rainfall tend to be most severe in areas where temperature or moisture is already near the limits of tolerance for a species. Thus, in the humid tropical lowlands where rainfall and temperature are high and near the optimum for the growth of the greatest number of plant species, there is relatively little annual fluctuation in either factor.

In the dry tropics not only is moisture naturally limiting for plants, but fluctuations in rainfall from year to year may be extreme. In northern temperate regions where winter cold and the duration of low temperatures restrict the growing season for plants, one finds the greatest variation between temperatures in one year and another. Climatic and atmospheric factors tend to have their most severe effects upon those introduced populations of plants and animals which man attempts to grow or produce in areas that have climates more extreme than the optimal climate for the species concerned. Native species are adapted to the range of climate in the area they occupy, and although they may be affected by climatic extremes, these effects are usually much less limiting than for a domesticated species introduced outside its normal climatic range. In the African savannas, a drought will be more likely to wipe out a high percentage of the cattle than of the native antelopes.

Climatic and atmospheric factors also have more effect of species that are at, or approaching, the level of subsistence density. Well-situated animals or plants, with adequate nutrients, water and other essentials, are less likely to be affected by drought than populations which have increased and expanded into marginal areas and are already in poor condition because of nutritional limitations. Certain factors of weather and climate may have catastrophic effects on all populations in an area—hurricanes, tornadoes, unusually severe floods or droughts, but usually these are less significant than the more normally occurring changes, except in the case of these species which are highly limited in distribution. With most species, normal fluctuations in climate have their greatest effect when animal population densities are excessive, or when plants are established in sub-optimum environments.

(b) Soils

Here the limiting factors operate directly on plants and, through plants, upon animal populations (other than those of the soil fauna itself, which is often directly limited by soil conditions). The requirements of plants and animals, however, are not identical. An abundant growth of vegetation does not necessarily mean there will be an abundant food supply for animals, nor that the food supply will have high nutritional value.

Certain plants may produce abundant vegetative growth and high amounts of carbohydrates under conditions where certain soil chemicals are in short supply; however, they may produce relatively small amounts of the proteins and vitamins needed to support an abundance of animal life.

Thus, a shortage of a trace element, such as cobalt, in the soil may have little effect on plant growth. However, cattle feeding on these plants may do poorly or not survive. Addition of a cobalt top-dressing to the soil which is then taken up by the plants can open up a rangeland to livestock use. For another example, a dense forest may produce an abundance of vegetation, but will support very few ground-dwelling grazing or browsing animals. There is no lack of food for them, but plants growing on a forest floor under dense tree cover are commonly lacking in protein.

Animals grazing of browsing them can receive adequate carbohydrates but not enough protein to thrive or reproduce. Thus, in many areas of dense forest, grazing and browsing animals will only be common around natural or man-made clearings. The abundance of natural vegetation in an area often has little relation to the inherent fertility of a soil. Thus, the ferasols and acrisols of the humid tropics are notoriously infertile, yet they support the most luxuriant vegetation to be found on earth. Examination of this situation reveals, however, that virtually all the nutritional elements required for the support of life are tied up in the vegetation and animal life of the area, and recirculate quickly from the dead plant or animal back into a new, living individual. Below the organic surface layer the soil contains few of the elements needed for plant or animal nutrition, and if the vegetation is scraped away with a bulldozer the soil that remains is often highly infertile.

By contrast, some desert soils that support little plant life, because of lack of water, may have a rich supply of the soluble mineral elements needed for the support of life. When water can be made available, they will support high levels of productivity, although there may be new ill effects such as salinization, waterlogging and water-borne diseases.

(c) Water

Water is an obvious limiting factor for plants and animals. Plants are classified in relation to their tolerance for dry, medium or wet conditions as *xerophytes*, *mesophytes* and *hydrophytes*. Most land plants are in the intermediate, mesophytic range; desert plants are usually, but not always, xerophytes; hydrophytes grow in water or require an abundance of water in the soil. All plants require water to support their active growth and metabolism.

Some plants cannot tolerate even brief periods of moisture deficiency and will wilt and die when the soil dries out. Xerophytes have special adaptations that permit them to survive under prolonged drought, but must have some moisture in order to grow and reproduce. Where present in abundance, water limits the occurrence of terrestrial organisms and provides a habitat for aquatic forms of life. The latter will relate in their diversity and abundance to the quality of the water, its ability to supply oxygen, its salinity, temperature, velocity and other factors. All of these are capable of being affected, usually adversely, by pollution or siltation.

In relation to terrestrial vegetation, an abundance of water or in the soil limits plant growth primarily by restricting the availability of oxygen. Those plants grow in wet soils or flooded areas have the ability to derive oxygen through special structure, such as devices that permit the aeration of their roots; the pneumatophores of mangroves, or the so-called 'knees' of balt cypresses are good examples. Continuously moist or flooded soils inhibit the bacterial or fungal action that would otherwise bring about decay of organic debris and litter. Organic remains will usually accumulate on such soils in the form of organic mucks or peat. Animals, like plants, require moisture for their continued existence, but vary greatly in their ability to endure drought.

Apart from purely aquatic animals, amphibious animals such as the crocodile and hippopotamus cannot long survive away from water. Most land animals must drink water regularly in order to continue to live. However, some desert species can exist indefinitely in the absence of free water as long as the food which is eaten contains sufficient moisture and they can find enough shelter to hold the loss of water from their body surfaces to a minimum. A severe drought, bringing an absolute shortage of water to an extensive area, can be fatal to all animals that require water for drinking. Nevertheless, much of the mortality attributed to drought results not from a shortage of water, but from a shortage of food within reach of water.

(d) Biotic Factors

The greatest number of limiting factors influencing plant or animal growth, abundance and distribution are biotic in nature. Food supply for animals is one of these, and is the most common factor limiting the growth of animal populations, either directly, through being short of requirements, or indirectly, through behavioural responses to food shortage. The number of plant-eating animals in any area is ultimately limited by the abundance of the plants of which the animals feed. Furthermore, since the plants must survive and reproduce, there must always be a greater abundance of plants than what is actually needed to provide food to the animals.

Most plants produce a surplus of vegetative growth and of seed, part of which can be used by animals.

However, each plant, in order to survive, must maintain a metabolic reserve—a minimum amount of leafage to permit it to store food for its own survival, or set seed for its own reproduction. Plant growth may be limited by competition from other plants of the same or different species each drawing on the same reserves of soil and water, or shading one another from essential sunlight. Some plants secret substances which inhibit the growth or establishment of other plants.

A variety of organisms may prey upon plants, from the seed stage through the life cycle to the mature plant. It is obvious that the number of carnivores or meat-eating animals in any area is limited by the availability of the prey upon which they feed. Also, there must be enough prey animals to enable the prey species to survive and reproduce; the predator cannot eat all of the prey, or both would become extinct. Most prey animals have developed behaviour patterns in relation to their habitat conditions that permit some of them to escape predators. Predators seek the most vulnerable and available prey, and usually do not hunt down and capture the more elusive individuals.

Often the individuals captures are either they very young and very old, or the most diseased or genetically least vigorous stock; therefore, natural predation tends to have overall beneficial effects on the population of a prey species. The relationship of the abundance of predators and prey is sometimes illustrated in the form of a *biotic pyramid*. Green plants occupy the base of the pyramid and must always have a greater total volume; *biomass* than the plant-eaters that feed upon them. The latter, the herbivores, form the second step in the pyramid, which must in term be wider and represent a greater biomass than the third step, representing the carnivores that eat other carnivores they must be fewer in number and contain less biomass than their prey. Each succeeding level in the biotic pyramid also represents a consistent loss of energy from green plants to herbivores to predators, since at each transfer of energy from one level to another some energy is necessarily lost. Far more calories are available at the lowest level of the pyramid than at any higher level.

The sequence of plants and animals feeding upon one another is called *food chain*. Food chains are interconnected, since more than one herbivore may feed on a plant and more than one kind of carnivore may eat a herbivore. These interconnected chains form a *food web.* As elements pass from soil (or atmosphere) to plants, and are then eaten by herbivores and in turn by carnivores, they are often concentrated in varying degrees by a process known as *biological magnification*. Thus, the element iodine may be present in the soil in small quantities and taken up by plants as it goes into solution and reaches plant roots. It has little direct function in plants, but is essential to plant-eating mammals. These must obtain enough iodine from the plants they eat and concentrate it in their thyroid glans for these glands to function properly.

The level of iodine in an animal's thyroid is far greater, therefore, than in either the plants or the soil. Although the number of herbivores is limited by the supply of plants, and the number of carnivores is determined by the numbers of their prey, the reverse is also true in varying degrees. The abundance and distribution of a plant can be limited by the presence of herbivores. If a plant is introduced into an area where there are no species that feed upon it, can spread rapidly and become transformed into a weed or pest.

Similarly, the abundance and distribution of plant-eating insects and other animals is often controlled by a wide variety of diseases, insects or other predators that feed upon them. If the predators are destroyed, their plant-eating prey can increase suddenly and explosively, since its reproductive rate in nature is adjusted to the need for survival under constant predation. Sudden increases in numbers of rodents or other plant-eating mammals have often been related to extermination of the predators that formerly held them in check.

Populations of many animal species are limited by the effects of interactions among individuals with in a species. The species concerned are described as *territorial* and comprise a proportion of individuals which persistently or seasonally occupy a particular area to the exclusion of other individuals of the same species. Commonly, territorial animals will tolerate or encourage the presence of certain other members of their species—a mate or mates, their juvenile offspring, and sometimes the members of a larger social group, a herd, troop, flock or pack. However, they will drive off or otherwise exclude animals that do not belong to the tolerated category.

Territorialism tends to limit the number of animals of any one species that occupy an area and to reserve for the territorial individual or group an adequate habitat and food supply; however, there are many kinds and degrees of territorial behaviour, and not all of them serve directly to limit population increase. Plants might also be considered to compete for 'territories' the well-established individual holds its ground against competitors. Their territorial defence mechanisms include the shading out of competitors and the secretion of chemicals which prevent rivals from germinating or growing. Numerous studies have now been made of the effect of overcrowding on a variety of animal species under conditions where food supply is abundant, and all other essentials (except space) are present in adequate supply. All of these studies show adverse effects upon individuals in a population where numbers exceed a certain level. These adverse effects are caused by social interactions, and vary from mild behavioural disturbances to serious forms of behavioural pathology, leading to increased mortality and a decline in birth rate.

(e) Interaction of Factors

Virtually all limiting factors interact and either reinforce or diminish their mutual effects Thus the effects of animal overcrowding may be seen in the limitations imposed by diminished food supply, increased predation, greater mortality from disease and various behavioural disturbances. Similarly among plants, factors of soil, water, air, nutrients, the presence or absence of intense sunlight, all interact to influence the germination or growth of a plant. Simple solutions to biological problems are therefore seldom likely to produce permanent, positive results.

Modifying the impact of one limiting factor may simply increase the operation of others. For example, in an area of scarce water supply it may seem sensible to make available additional water in order to permit animal populations to increase. Yet doing so has, in many observed instances, quickly led to major die-offs from malnutrition resulting from an insufficient food-supply for the increased population. Similarly, destruction of predators may cause protected herbivores to die in an epidemic and the elimination of a disease may cause a crash die-off from starvation. It is therefore always necessary in management planning to look at the operation of the total ecosystem and consider how all the factors interact. Failure to recognize these principles has caused many economic development activities to fail and much money to be wasted.

Diversity, in this context, refers to the number of species of plants or animals that live in a particular area.

It has long been recognized that biotic communities found in more extreme climatic (and geological) environments are relatively simplified and less diverse than those occurring in climatic regions more consistently favourable to the growth and survival of living things. Thus, the cold subpolar regions of the world support relatively few species of plants and animals, the same being true of arid deserts, high mountains and the slopes of recently active volcanoes.

The greater diversity of plant and animal species is found in the humid tropical regions of the earth, where climate is relatively stable, uniform and favourable to plant life. In the more simplified natural communities, the lack of species variety is matched by the frequent occurrence of large numbers of a single species: the great herds of caribou or reindeer in undisturbed areas of Arctic, or the enormous colonies of penguins in the Antarctic are good examples. By contrast, such relatively high numbers of a single species are not commonly observed in the humid tropics, except where special edaphic, physiographic or successional conditions prevail, as in the case of mangrove belts in estuarine areas. Indeed the rarity of big stands of a single valuable timber species has been a factor militating against long-term economic management of tropical rainforests. In less diversified natural communities there are also great fluctuations in populations of some animal species and annual plants from year to year.

The cyclic increases and declines in the numbers of arctic lemmings are a classic example. Similar fluctuations occur in many other arctic species. In desert regions, population fluctuations are not as regular, but may be even more severe; examples are the locust plagues and the massive increases in numbers, followed by major die-offs, of rodents or hares. It has been concluded by many ecologists that diverse natural communities, because of the more numerous interactions among species, are more stable than simple communities. Furthermore, such communities are more productive in a gross, biological sense although the gains in organic matter production are generally off-set by losses through increased respiration. When these communities are simplified by human intervention, energy is channelled along more limited pathways and the species that remain may grow rapidly and achieve high levels of net productivity.

However, because the normal buffering actions, or checks and balances, present in diverse communities have been removed, such communities are comparatively unstable. High net gains in organic production by one species in one year may be balanced by major losses in another. Management of the human environment has generally taken the form of simplifying complex biotic communities, removing those elements less immediately useful to man, and substituting species more obviously useful. A complex rainforest is reduced or refined to a few valuable species or replaced by a simplified plantation forest; a grassland is converted into a grain field, and so on. By such means, man has managed to obtain high yields of species he favours. But these yields are likely to be either temporary or highly fluctuating, unless factors which, in a natural community, lead to stability and productivity are compensated for by natural community.

Biotic succession and Land Rehabilitation

The ability of an ecosystem to tolerate human use and to recuperate from human abuse varies greatly with climate and biological factors. The recovery of communities from fire illustrates this. In areas that usually have high rainfall and humidity, for example, naturally occurring fires are rare. In unusually dry areas, however, large fires may burn. The immediate results can be disastrous. However, the results of the fire are seldom long lasting. A process known as biotic succession begins, and follows a more or less regular and predictable sequence. In forest areas a weed stage commonly takes over first on burned ground. After a few years, shrubs and trees of species that germinate on open, unshaded ground will move in and shade out the weeds.

The early shrubs will be replaced by the light-tolerant or light demanding trees. These in turn provide cover for those shade-tolerant species of trees that can seldom establish themselves in the open. As these grow, they will replace the shade-intolerant species. Eventually, and provided a seed supply is relatively near, the species characteristic of the original forest will occupy the ground; and its structure, physiognomy and botanical composition may remain stable for an indefinite period of time. This relatively permanent form of vegetation is known as climax.

The vegetation that precedes it in occupying the open ground is successional. Succession can be diverted or delayed. Thus, if bare ground in the tropics is invaded by tall grasses of such species as Imperata cylindrica, they may hold the ground and resist invasion by trees and shrubs for some time. If the grasslands are burned frequently, invading trees and shrubs can be indefinitely excluded. As so-called 'disturbance' climax or 'fire' climax may be the result. Other disturbing factors produce results similar to those brought about by fire.

Grazing of climax grasslands that have not recently supported grazing animals results in a shift in the species composition of the vegetation towards those plants best able to withstand cropping. If grazing is removed the less tolerant plants will recover, but if grazing pressure is increased to high levels virtually all vegetation may be killed. However, recovery will usually take place along predictable successional pathways if the grazing animals are reduced in numbers or removed entirely. Similarly, land clearing or any other process that removes the original vegetation will set in motion successional processes which will eventually lead towards recovery of the climax. The climax vegetation is that which is best suited to the climate, soils, physiography and biotic influences of a region. Since biotic succession leads towards this climax it is difficult, without constant management, to maintain an area in vegetation which is successional by nature or is exotic to the area concerned.

The continuing battle against weeds on agricultural ground is a battle to arrest normal successional processes. The efforts to keep savanna grasslands free from woody vegetation represent a similar struggle against succession. It should be added that the rates of succession will vary, of course, with soil fertility. If the latter is lost, there will be, at best, a very slow rate of establishment and succession of new vegetation.

Resilience of Populations and Communities

The ability of a community to recover from disturbance is a measure of its resilience. Some ecosystems are highly resilient, others are exceedingly fragile. As noted above, communities of arid or cold regions may be fragile, and their ability to recover from disturbance is limited. Communities in more favourable environments tend to the resilient. There are, however, exceptions to this rule, such as the highly specialized communities which have developed on some isolated island and high mountains.

Islands which are located at considerable distances from the mainland can be reached only by those species which posses considerable powers of dispersal: plants with light, wind-borne seeds or spores, micro-organisms with resistant eggs or spores, flying animals, species that can withstand ocean voyages on floating rafts of natural vegetation, plants with water-resistant seeds, etc. Most mainland species cannot reach distant islands, and in consequence, islands have developed unique, specialized biotic communities.

Considerable internal evolution has taken place on islands which have been isolated over millions of years, such as the Hawaiian chain or the Galapagos. Entirely new species, not found on the mainland, have evolved. Island communities, therefore, are highly susceptible to disturbance. Where man has introduced mainland species of plants or animals, particularly very adaptable ones, these have often spread through the island communities, displacing the native species which evolved in the absence of such competition from their mainland relatives. Island animals may have developed no behavioural responses to predators, if predators have not evolved on or previously reached the islands. In some cases they may be entirely 'fairless' in behaviour, and some island birds have even lost the power of flight. Introduced predators— rats, cats, mongooses, etc. and man himself--can therefore easily cause their extinction. Island plants may have little resistance to grazing use and may be invaded and native species displaced by vigorous species from the mainland, as has happened with *Lantana camera* in Hawaii.

Attempts to develop island environments can have disastrous effects if they lead to decimation of native flora and fauna, which is not only an important scientific and recreational loss in itself, but can also lead to accelerated soil erosion, breakdown in soil fertility, and total disruption of island ecosystems. High mountain tops represent environments which differ markedly from those of the lower slopes. Alpine areas in the tropics often have more in common with arctic regions than with the humid tropical lowlands. Where high mountains are isolated from one another, like Kilimanjaro, Mount Kenya and Mount Elgon in East Africa, their upper altitudes have the same degree of isolation as oceanic islands.

Such areas can be reached only by the more mobile animals adapted to their climates and vegetation, and by plant species whose seeds can be dispersed by wind or carried by birds. Like islands they become centres for local evolution and, like those of islands, their plant and animal communities can be highly fragile, because of their isolation and the rigorous climatic conditions under which they live. An equivalent care must be exercised in their management if their productivity, usefulness and biological values are not to be destroyed.

Ecological Balances and Survival Thresholds

Any ecosystem, no matter how resilient, can be pushed to a point of no return or, more exactly, to a threshold beyond which limiting factors become so severely operative that recovery, in periods meaningful in the human time-scale become impossible. For example, soil erosion on mountain sides may reach a point where bedrock is exposed. When this occurs only the slow process of primary biotic succession, operating centuries or millenia, can build the soil and vegetation.

Another threshold process, typical of many areas, is soil laterization, which can only be arrested economically at certain early stages. Once a tough layer of laterite is exposed or formed, recovery may no longer be feasible. Communities can repair themselves, with moderate protection, up to a certain point. Beyond level of abuse or disturbance biological repair becomes, in human terms, an intolerably slow process.

Consistent overcropping of any species can reduce its number to levels from which recovery is no longer possible. Killing the last individual animal or cutting the last tree of a species may not be needed to bring about ex-termination. If animal populations are reduced to a level at which effective breeding does not occur, or if the habitat of a species is destroyed, recovery will not take place, even though some individuals may for a time survive.

Similarly for plants, if a minimum area to ensure regeneration of certain species is not maintained, they will eventually die out. The number of individuals or the size of habitat needed to assure survival of species has not yet been generally determined. The threshold apparently varies with the genetic make-up of the species, its habitat requirements and behavioural characteristics.

Preservation of Environmental Quality

Up to the advent of civilized man, the biosphere of the world consisted of a complex pattern of biotic communities responsive to a variety of environments. The harmony prevailing between the biotic and physical realms was the result of hundreds of million of years of evolution and adjustment of one organism to another and of each to the habitat in which it lived. The relationship was a salient one.

A sporadic outbreak of a species as the result of some temporary advantage, or a catastrophe to a population as the result of a change in the physical environment, soon became tempered and a balance restored. Primitive man occupied a niche and assumed a role comparable to other animals and the ecosystem maintained itself in a vigorous healthy condition. Modern man, however, has initiated a number of abrupt changes during the last few centuries beyond the capacity of the ecosystem to adjust. He has become the dominant organism in the biosphere, forcing changes to fit his needs and desires, and creating ecosystems different from any that have existed before. Emergence of new communities is a characteristic of the geosere, but the replacement of the biosphere by an "anthroposphere" or "noosphere" (from the Greek word, noos, mind) is unique in the speed and extreme changes produced. This acquisition of dominance is the result of the extraordinary evaluation in man of intelligence, ingenuity, and use of tools. He has greatly modified or replaced the natural processes that control stability and balance within the ecosystem with new processes and manufactured products which the environment cannot absorb.

The result has been a general deterioration of the environment, a population-explosion of the human species, and a disharmony in the world ecosystem and will, unless resolved, lead quickly to catastrophe. Whether man will survive as the dominant organism, be reduced again to the role he occupied as a primitive, or become extinct may well be decided within the next few decades. There is conflict here between the demands of man for expanding population, productivity, power and pleasures and the functioning of natural laws for maintaining stability and equilibrium between the living and physical environments. Immediate problems of concern in the preservation of natural environmental quality are the elimination of air, water, and soil pollution, the disposal of solid wastes; and climate modification on a planetary scale. Perhaps noise pollution should also be added.

Pollution is change in environment, caused by man's discharge of matter or energy into it, which alters the normal function of an ecosystem. Fundamental to the attainment of stabilized and balance worldwide ecosystem are the control and regulation of the human population at a suitable level and preserving the habitat and the basic community of plant and animals life upon which human life depends. Considerable technical knowledge already exists for solving these problems. The immediate need is concerted attention for implementation of the necessary remedial measures. Satisfactory resolution of the problems will involve not only science and technology but also law, sociology, politics and economics. Emphasis on existing knowledge, however, should not obscure the fact that continuing basic research is required to elevate man's understanding of the environmental system.

6 MAN'S FUTURE

The most difficult and essential problems for man's future lie in broad ecologic and social issues. No two questions are more fundamental or more troublesome to the future of man than the ecologists' inquiry. "What is the carrying capacity of the earth for man? In other words, what are the limits of the life support capabilities of the earth in terms of human population?" are the behaviorists' concern. "What are the tolerance levels of man to the pressures generated by his own societies? In other words, if man is his own worst enemy, does this lead to his self destruction?"

The carrying capacity, or life support capability, of the earth for man may be dependent upon a great variety of factors. We can readily identify oxygen water, food space and shelter as vital requirements for human life. There might be many others as well such as mineral resources, energy availability, or the concentration of toxic metabolites, which could reach states above human tolerance levels. Most of these however, would operate through the media of air, water or food. Hence, it is worth while to examine each of these briefly.

Oxygen supply and Air Quality

The atmosphere of the earth consists of about 78 percent nitrogen, 21 percent oxygen, 0.03 percent carbon dioxide, and the remainder, other gases. This composition is remarkably stable despite the fact that oxygen is a highly reactive element which readily combines with a great many other elements in chemical processes of oxidation. The ecologist Lamont C Cole has pointed out that oxygen has remained stable in the earth's atmosphere because green plants recycle it in molecular form through photosynthetic process.

Approximately 70 percent of the free oxygen produced each year is through to come from planktonic plants in the ocean, and the remaining 30 percent from terrestrial vegetation of forests and grasslands. Cole emphasized that if we should seriously pollute the sea, as is already happening in coastal areas, and if we should excessively denude the forests and grasslands of the earth, we shall be in danger of upsetting the oxygen balance in the atmosphere. In other words, we must look upon the oceans, forests and grasslands, as the world's oxygen tank, and we must insure their continued healthy ecologic function. This point of view is contested by W.S.Broecker (1970), a geochemist at Columbia University Dr. Broecker points out that each square meter of the earth's surface is convered by 60,000 moles of oxygen gas. Plants in the ocean and land produce only about 8 moles per square meter per year, an this is almost entirely consumed by their own respiration and that of animals and bacteria. Thus other organisms use all but 0.0004 percent of the oxygen produced by plants.

Broecker maintains, therefore, that man does not depend upon the oxygen generated by plants, and that there is no current danger at all to his oxygen supply. "In conclusion it can be stated with some confidence that the molecular oxygen supply in the atmosphere and the broad expanse of open ocean are not threatened by man's activities in the foreseeable future. Molecular oxygen is one resource that is virtually unlimited". Broecker feels that man's existence will be threatened by other types of environmental pollution long before his oxygen supply becomes endangered, and that we need not worry about total oxygen supply. This is typical of some of the different opinions which are unfolding between physical and biological scientists. Such differences often indicate the need for more research and better knowledge before we can fully accept one position on another.

Another atmospheric problem of long-term concern is the increase in carbon dioxide in the atmosphere. Between 1860 and 1960 the combustion of fuel increased the CO_2 content of the air by 14 percent of its formerly stable level, Carbon dioxide tends to reflect infrared rays from the earth, and this provides a "greenhouse effect" for the whole earth; that is, it accumulates heat. This will gradually increase the temperature of the earth—an effect which may have long-term implications in climate modification.

A rise of just 2 or 3 degrees in annual mean temperature would have a major melting effect on glaciers and ice caps, and this could raise the level of the seas several feet. Such a rise in the seal level would inundate many of the world's major cities. Some experts predict it may require 4,000 years, and others feel it will not happen at all. The latter believe that any occurrence of the greenhouse effect due to increasing CO_2 in the atmosphere will be offset by increasing particulate pollution in the air and a cooling of the earth's surface. Once gain, we see fundamental controversies on major issues-largely because we do not really have the complete answer. The atmospheric problem may also become acute in terms of various toxic elements released into surface, air and water: carbon monoxide, sulfur dioxide, lead from automobile exhaust, hydrocarbons from industrial operations and numerous other products of modern technological societies. For example, 500 million pounds of lead are discharged into the atmosphere of the United States each year from automobile exhausts alone, and significant amounts of this become incorporated into living organisms.

A recent study in Sweden showed concentrations of lead in mosses to be as high as 300 to 500 ppm along roadsides. Some roadside plants in the United States may have lead levels up to 3,000 ppm. This lead enters the plants via both air and water and can then enter man and animals. A hundred years ago, before the automobile, concentrations in roadside plants were less than 20 ppm. Lead levels over 0.5 ppm have detectable physiological effects in man and animals, and 1.0 ppm levels are often toxic.

There seems to be little or no immediate threat that we will run out of air for breathing, but several warnings have appeared regarding the quality of our air. We apparently have an adequate supply of total oxygen for many centuries to come though even this is a debatable point. In some localities the situation may be critical at certain times. If stationary inversions of air trap the polluted atmosphere over some of our largest cities such as Tokyo, London, New York or Los Angels, toxic air pollutants increase to dangerous levels. In August of 1969, a massive still-air inversion occurred in eastern United States, causing serious air pollution in urban areas from New Orleans and St. Louis, north to Chicago and Milwaukee, and east to Washington and Baltimore.

At no time was there an absolute shortage of oxygen, but dangerous level of carbon monoxide, sulfur dioxide, hydrocarbons and other air pollutants accumulated in several cities. Twenty million people were affected by varying degrees of eye irritation and respiratory diseases. Fortunately, this great blanket of smog was dispersed by upper air currents in the first week of September, and no large-scale health tragedies occurred. More serious event of this type are likely to occur in urban areas unless strong.measures against air pollution are undertaken.

Water Supplies

Worldwide water supplies represent a concern for the future of man in terms of both the quality and absolute quantity of fresh water. Many cities have reached or are coming close to their total available fresh water supply and are reaching further and further to obtain new supplies. Los Angeles, for example, is planning tremendous canals and canduits which will bring water from the pacific northwest over 1,000 miles to the growing megalopolis of southern California. Two Russian geographers and hydrologists, Kalinin and Bykov, have pointed out that "The world's water resources from a single entity.

The basic hydrologic cycle of the world depends upon solar energy, atmospheric circulation, heat reflectivity and surface retention. The latter is strongly influenced by the biotic community, kalinin and Bykov point out, on the assumption that the moisture content of the earth remains constant that the total gains and losses from the land and the sea are in balance. Thus, the total precipitation equals the total evaporative loss from land and sea. However, the relative retention time and thus the total amount of water in terrestrial and fresh-water environments can change substantially in relation to the biota and terrestrial storage capacity. A large proportion of the world's fresh water, falls and resides in relatively unpopulated portions of the globe, particularly Siberia and northern Canada. Here, man's activities are limited by other factors and he cannot readily obtain these vast stores of water. Secondly, the matter of water quality is of immediate concern. The principle of water reuse is now becoming accepted, so that people recognize much of their water has been used previously by other municipal systems. Each reuse cycle however, is increasingly costly in terms of treatment, and each reuse cycle requires disinfectants and other industrial chemicals, combining these problems of distribution and quality, it seems that the water problem of the future is not going to be one of a global water shortage on a planetary basis but one of local crises due to the concentrations of people, industry and water use.

Certainly, the immediate concern is finding local water resources of adequate quality to meet the burgeoning needs of urban populations. To the city dweller in New York or Hong Kong whose water tap runs dry or to the refugee in the Middle East who is dying of thirst, it well be little comfort to realise that 29,000,000 cubic kilometers of fresh water are locked in glaciers and polar ice caps.

World Food Supplies

Nowhere is scientific controversy more evident than on the issue of world food supplies. Some scientists predict famines and mass starvation within the next 10 years. They feel the stage is irrevocably set for millions of people to die of starvation in the 1980's. In fact, we already have famine in many parts of the world, if one defines famine as people dying of lack of food. Even at the height of the Green Revolution in Asia in which new strains of rice and wheat were producing phenomenally large yields, famine conditions existed in parts of Rajasthan in India: in sections of Burma and in southern Thailand.

Starvation is now a grim reality in much of the world and it may well get worse despite the Green revolution. Other scientists refute the pessimism and find evidence for a hopeful outlook in the world food situation. They feel that the Green Revolution which was based on new strains of rice and wheat in Asia and Africa, and which produced a doubling and tripling of yields in one or two years, shows what can be done with modern agriculture.

For the first time in many decades, agricultural output increased more rapidly than population growth. India began to approach self-sufficiency in food production, and the Philippines began to export rice again for the first time since 1900. A word of caution about the Green revolution is also necessary. It is quite true that new strains of agricultural crops can sometimes dramatically increase yields, but often these new strains require more careful management, more intensive care, more water, more fertilizer and more insecticides. They may be more vulnerable to plant viruses or fungi.

Although the plant geneticists try to breed in every possible disease resistance, these new strains do not have the centuries of natural selection and regional adaptation behind them as do the local strains which the villages have always used. In West Bengal and Orissa, there were reports that some of the newly introduced high-yield strains of rice were succumbing to a disease that was not affecting that old traditional strains. If this happens to a major extent, the Green Revolution could collapse and a severe blow would be dealt to the villager's faith in the agricultural scientist. Another factor in the world food picture is one of pure climatic good fortune.

Green Revolution in India was accompanied by the best monsoon seasons in over 10 years. A long series of droughts had broken is southern Asia and adequate rains fell in the right pattern. The possibility exists, therefore, that a good part of the success of the Green Revolution was due to good weather, combined with new strains and more fertilizer. This is a question that will be more adequately answered in future year to see if increasing agricultural production can be maintained in years of poor rainfall. In India too much rain came and it fell to early. The wheat crops in northern India were magnificent in April and May just before harvesting and threshing, but heavy premature rains in many areas caused the wheat to rot before it could be threshed. Bihar had serious crop losses from flooding. Once again, this was a tragic and dramatic demonstration that even the Green Revolution remains dependent on climatic factors outside the control of man.

Through all of these arguments on how many people the world can feed, one must ask the realistic question: "Is it necessary that there always be a race between population and food production? Is it desirable to be teetering on the brink of starvation at all times? Is it the best representation of the human condition to be struggling in this manner to feed more and more people at a subsistence level? One also has the impression in reviewing the literature on world food supplies that too much emphasis is placed on future predictions, and not enough on current conditions. We furthermore know that malnutrition exist throughout the world, even in the most advanced countries. We furthermore know that malnutrition of many kind has profound effects upon growth, development and behavior.

Children growing up with malnutrition from early infancy definitely show reduced growth rates and possible retardation of physical and mental development. They furthermore show increased susceptibility to infectious disease. Thus, we return to the issue of the quality of life, and the realization that merely preventing death by starvation does not by any means insure a quality nutritional base for adequate growth and human development. It is apparent that the problem of world food supplies will continue for some time as a problem of both quantity, and quality and especially one of distribution.

Space on Earth

In 1958 Paul Sears wrote an important essay entitled, "The Inexorable Problem of Space," in which he concluded that "Our future security may depend less upon priority in exploring outer space that upon our wisdom in managing the space in which we live". The earth's total surface is 196,940,400 square miles of which about 30 per cent or 57,230,000 square miles is land surface. Of this total land surface, approximately 5,000,000 square miles are in Antarctica deserts, and 11,000,000 square miles are in mountainous regions. Of deserts and moutains, at least one-fourth of their jano area cannot be inhabited, so this eliminates another 5,000,000 square miles of land surface for human habitation.

We are left with approximately 47,000,000 square miles of land surface, at the most, for the world's population. The world now has nearly about 6 billion people. The world population has doubled in the past 40 years and may double again in the next century, perhaps approaching stability at about 11 billion by the year 2100. By 1990, of the world's 5.3 billion people, 4.1 billion—77 percent lived in the developing world; 1.2 billion inhabited the industrialized countries. India has nearly 1/6th of the world's population. There are 257 person per sq. km. in India. In the Capital City Delhi there are 6318 person per sq. km. Thus it is to say human population is not evenly distributed. It leads to the problem of space.

As with water, air and food, it is important to distinguish between total or potential amount of any given resource on earth and that amount which is actually utilized or practically available to man, it was pointed out in the sections on water and air that the earth's total amount is apparently adequate for many years to come, but the supply in any given city may not be of sufficient quantity or quality to support the standard of living we would like. The amount of living space per person in some cities is less than 20 square feet per person.

Theoretically, we could redistribute these crowded people to provide more even spacing around the world. Obviously, this cannot be done, and if it were attempted it might wreak havoc with agricultural production. The economic and behavioral needs of man are such that he must cluster together, but in so doing, he creates superdensities such as Calcutta, London, New York, Paris, Tokyo, Shanghai and Buenos Aires. The slums of which are definitely deleterious to human health and welfare. Why, we might ask, does man insist on this degree of crowding and then complain about it so bitterly? Why do people continue to pour into the cities of the world when they are already known as centers of population crime and illness?

Perhaps we need to know more about the real spatial and social needs of man. What are his requirements for space and privacy, and how do these interact with his needs for social stimulation and group activity? Does man become conditioned to crowding in early childhood and then seek further crowding throughout his life? Does he require the social pressure of a crowd because he feels a compulsion to be in touch with the group and thus be able to predict what the group is going to do next? Have we produced, in our urban centers, self-destructive, systems which so modify the behavior of their inhabitants that these inhabitants exaggerate their own problems and can no longer achieve rational solutions?

These are all tough question that cannot be answered in simple or easy terms. The answers will require the best intelligence we can muster from many disciplines. Some sorts of answers will certainly be necessary before we can even begin to solve the ecologic and behavioral problems associated with our own use of terrestrial space.

Distributional Inequality

With all of the last four life-support requirements of man-air, water, food and space—we have seen that the major problem of immediate importance lie in distribution more than in total amount. Technologist therefore look upon these problems as soluble by technological means. They feel, they are essentially matters of economics and transportation Behavioral and environmental scientists see them as more basic, however, relating to man's fundamental ecologic behavior. They feel, they are essentially matters of economics and transportation. Behavioral and environmental scientists see them as more basic, however, relating to man's fundamental ecologic behavior. They feel that certain basic social and ecologic tendencies of man prevent equitable distributions of the earth's resources. Obviously this type of argument can go on forever, and it is perhaps necessary to take a pragmatic approach and look at the current picture for the best glimpse of the future.

Many of the industrial nations of the world, especially the United States, the national northern Europe and Japan, have enjoyed rising prosperity for the last 25 years. In ecologic time, of course, this is but a flash in the pan, but let us assume that it can be sustained for another 25 or 50 years. It is worth examining where the material and energetic support for this prosperity arises. This distributional inequality is another major symptom of ecologic instability in the world.

One would hope for some narrowing in the gap between the haves and the have-nots, but unfortunately the opposite trend is still occurring. The wealthy nations of the world are far outstripping the poor nations in their rates of economic gain and human betterment. There are, however, self-corrective mechanisms within this system as, for example, when some segment of the labour it can find which will do a satisfactory job. Thus, there has been a great expansion of manufacturing in Japan, Hong Kong, Taiwan and Singapore in industries as diverse as steel, electronics, textiles, toys and optics.

The Western factory worker experiences for the first time an improving situation. The western industrial worker continues to refuse his own working prospects by striking for higher and higher wages. This will certainly be one mechanism by which wealth is more adequately balanced between nations. Still the mass of poverty throughout much of Asia, Africa and Latin America is so great that it is difficult to imagine a painless equilibrium. The differential distribution of wealth remains sharp and represents a type of ecologic situation which cannot last indefinitely without severe consequences and reverberations. To the ecologist it means either a successional stage which is not yet mature, or a system headed for a major breakdown.

Economic and ecologic maturity should ideally be in the direction of broader and more equitable distribution of the earth's resources—a narrowing of the gap between the haves and the have-nots, both within and between nations, provided the total system does not pass beyond the state of corrective change.

Violence and War

Any ecological system or biotic community can be overloaded to the point of collapse by a variety of factors. Certainly one of the factors for man of most immediate and realistic danger is that of social violence. Either by war or by civil riot, man threatens his own existence. To maintain a sense of physical security man has built an incredible stockpile of destructive power. The nations of the world spend more than 200 billion dollars per year on armaments—more than on health education, welfare, agriculture, child care and care for the aged combined.

Twenty tons of explosives are stored in the world for every man, women and child. Russia and the United States have stockpiled 6,000 nuclear war heads possessing thousands of megatons of explosive force. Each megaton has more destructive power than which existed in any of the previous wars of history. The world spends almost 8,000 dollars per year training each soldier and less than 100 dollars per year educating each child. The question is, of course, how long can man continue to preside over armament like this without using it? And the answer seems to be, not very long. The fact is we do use part of the power in an attempt to achieve social goals. It is being used daily in various parts of the world to maintain territory and dominance ranks. It is justified as a spare and deliberate use in hopes of avoiding the all out effort. How long this can continue is anyone's guess, and the ecologist can claim no more omniscience than the political scientist.

The ecologist points out, however, that the ecologic and social pressures which generate international military tension may rise with increased population pressure. It is not likely that man can solve the problems of war and violence until he approaches and understands their basic orgins in terms of the behavioral and ecologic relationship between social groups, and the environment.

Adaptive Capabilities of Man

Many scholars feel that one of the great problems of the present and the future is the rapid rate of change which is occurring in society and technology. There is not, however, agreement on the consequences of the rapid change, changing life patterns are too rapid for man to adapt successfully. Mankind is so overstimulated by the increase in new sight and sounds, so bombarded by world wide mass media, and so constantly beset with noval situations and new problems that he will break down psychologically and physically. His responses will be maladaptive and pathologic. He will thus collapse under sensory overload. It is quite possible that each of these points of view has large elements of truth in it. In the present world one can find many examples of both.

Our mental institutions are already filled with those who could not adapt to sensory overload and who experience a personal breakdown of physical and mental health. Yet one can find people are living a sub-human existence where the quality of life is so low that the amount of human pain and misery seem unbearable. Yet man has adapted successfully to these inhuman conditions. In many ways, the future is with us now. We can find throughout the world an abundance of high quality life in high quality environments, but we can also find an abundance of low quality life in impoverished environments. Many of the questions about the future revolve, therefore, around the question of relative balance and proportions. Which way are we heading and which way will the scales tip?

A major aspect of the future centers around our sense of responsibility to posterity. No representative of posterity is here to speak for the segment of society. How do the present populations of mankind regard their obligations to future generations? Are we willing to be more conservative in our life requirements, and are we, in fact, willing to reduce our rates of personal economic growth to insure a better environment and a greater abundance of resources for future generations? This may be one of the central issues of the next twenty years, and one on which the quality of present and future environments depends.

World-Wide Environmental Issues

Environmental problems have become a global issue: Some of the world wide issues are briefly described below:

(a) Climatic changes/Green house effect

Many studies show that the atmospheric pollution is on a global scale. Today, however, it is not only the air over our cities that is polluted, the entire atmosphere of our planet is now afflicted to some degree. Meteorologists talk about a nebulous veil of air pollution encircling the entire earth. Smong has been observed over the North Pole, and in their unlikely places.

There are range of environmental pollution issues originating largely in developed countries which are also of global significance. Carbon dioxide (CO_2) build-up leading to climate change is one of those key issues. Recent research indicates that increasing concentrations of CO_2 in the atmosphere, due mainly to the burning of fossil fuels could cause a warming trend leading to climate change in the next century. While there is still scientific uncertainty about the pace and impact of such a change it could be of sufficient magnitude to produce major physical, economic and social dislocations on a global scale.

A special challenge posed by the issue is that if preventive actions are postponed until some effects are perceived, it may be too late to avoid appreciable climate change. It is therefore in the interests of all nations to support international efforts to understand the phenomenon at early enough stage for action to be taken to reduce or avoid major economic, social and environmental disruptions around the world. However, as developed countries are currently responsible for over two thirds of total world emissions of CO_2 by fossil fuel combustion, they have a special responsibility for supporting and contributing to accelerated efforts to assess more precisely the nature of the problems and the alternatives for action to reduce or avoid any serious risks. Another issue is the world-wide distribution and use of harmful or potentially harmful chemicals.

There are now some 70,000 chemicals on the commercial market, and many are currently used and released into the environment with little or no knowledge of their potential long-term effects. It is estimated that another 1,000 new chemicals enter the market each year. Chemicals cross national boundaries in a variety of ways: they are transported by winds and water.

(b) Risks to Ozone Layer

Ozone is a different form of life sustaining oxygen, having three atoms instead of the usual two. But this third atom, scientists have found, is very unstable and can be eliminated by chlorine ions. The scientist have now found a close link between the family foam mattress, refrigerator and the air conditioning system of an average home and Ozone layer, 50 to 70 kilometers above the earth. They believe that free chlorine ions into the atmosphere, from plastics and other chemicals remove that extra atom of oxygen which makes Ozone. Only, in this case, the loss of the one atom changes the ability of the layer to act as a strainer to absorb the damaging ultraviolet B wavelengths from the sunlight.

The findings by the experts have revealed that the rate of Ozone Layer depletion is sufficiently large to present serious consequences for man, animals and plants. In addition, new threats to the ozone layer are being recognized.

Take, for example, the refrigeration industry. Food preservation has become a necessity and refrigeration is here to stay. Only advanced technology can provide the solution to prevent the release of the damaging ions. Another example is nitrogenous fertilizer which, it is suspected, release chemicals damaging to the Ozone layer. The demand for more food and the booming green revolution in many developing countries have pushed up the use of this fertilizer. Today the world uses about 50 megatons of it. In another 20 years the rate of uses will be 100-200 megatons.

Though the Ozone Layer accounts for less than one millionth part of the earth's atmosphere, it is of vital importance to the earth's life form. It has two main functions: first it absorbs the UV-B while allowing the heat generating infra-red rays to reach the earth freely. The heat from infrared rays sustains plants and human lives by keeping the temperature of the planet at a desired level. Secondly, it influences the whether by stabilizing the stratosphere which acts as a cap to the turbulent weather system in the troposphere. Scientists knowledge of the Ozone layer is still limited for want of advanced technology.

The presence of other substances in the stratosphere, which absorbs UV-B, makes it difficult to come to a numerical and logical conclusion. But the scientists do agree that an average 70 per cent of the total UV-B reaching this planet falls on the equatorial regions between the latitudes 30° S and 30° N. They also agree that for genetic reasons people with light eyes and light skin (including albinos) are most affected by UV-B. Scientists today claim UV-B can destroy micro-organisms outright as well as cells in plants and animals. It also destroys the DN molecules and the genetic cell composition.

Under natural conditions plants and animals have

developed behavioural adaptation but gradual or sudden increase in UV radiation due to changes in the Ozone Layer is likely to disturb this balance. While the scientists agree that UV-B contributes to the induction of cancer they agree the dose response relationships are too complex and unknown. These make the quantitative analysis speculative, as regards the cause of melanoma cancer and resulting deaths.

The North European Conference of Scientists November, 1980, warned that non-melanoma skin cancer in light-skinned people would increase and many crops and many acquatic organisms, including shrimps ad fish eggs, could be damaged by increased UV-B infiltration. This could even affect the reproductive process of plants and animals. It recommended that existing satellite and ground based ozone observing system should be integrated through Global Ozone Observing System (GOOS). They also recommended updating of equipment and setting up of more stations for observing the ozone system.

The scientists concerned with outer limits of the biosphere felt they are not yet technically equipped to measure accurately the rate of ozone depletion-already depleted by one percent. Though the danger is known, science has not yet provided a perfect model of measuring the depletion rate.

(c) Acid Rain

Every one realises how important the rain is. Without it the continents would be barren. Yet now the rain in parts of the earth has taken on a new and threatening complexity. It mixes in the air with pollution from burning fossil fuels particularly in power stations, factories and motor vehicles and brings down dilute sulphuric acid. This is killing fish and water life, and corroding buildings, including some of the world's most important ancient monuments. It may also damage forest and croplands, and possibly pose a substantial threat to health.

Acid rain is now recognized as an international problem. The air of industrial towns partly is cleared by chimneys at factories which push the pollution high into the air. These chimneys have made the things better locally by dispersing the pollutants, but aggravated the international difficulties. For the sulphur and nitrogen compounds emitted by burning fossil fuels can be blown thousands of kilometers by the winds, to cause acid rain in countries far from their points of origin.

A special conference on acidification of the environment was held in Stockholm in 1982. It reviewed and assessed a large amount of scientific information not previously available. So far, acid rain has been seen as an essentially regional problem, confined to the industrial areas of the northern hemisphere. But, though the problem was first perceived there, it may become much more widespread for acid rain is likely to occur wherever fossil fuels are intensively used. Acidification is an environmental problem, or becoming one, in parts of Europe and North America. Similarly polluted areas are likely to exist elsewhere in the world, especially around large urban and industrial coagulumerations.

Soils are normally much better able to resist acidification than lakes, rivers and streams. Meanwhile, acid accelerates corrosion in most materials used in construction of historic buildings, and other important cultural objects. Some of th worlds greatest cultural treasures including Parthenon in Athens and Trojans' column in Rome are being eaten away by acid fall out. There must be more effort to research and develop ways of improving energy conservation, environmentally appropriate technologies for producing heat and power, and techniques for removing sulphur from fossil fuels and gaseous emissions. The problem of acidification may well spread to new areas as a result of rapid industrialization and the growth of cities in other parts of the world, particularly developing countries. So it is important that areas affected by acid deposition and susceptible to damage from acidification are identified as soon as possible. It they are, the damage could be mitigated or even avoided of a minimum cost to society by initiating research and applying what we already know about pollution control.

(d) Pollution of World's Oceans

The data collected from the Baltic, the North Sea, and some North American and Australian stuaries and coastal waters suggest that in most of these areas the pollution that caused most concern was due to sewage, agricultural chemicals, oil and metals. Metal concentrations were clearly elevated in coastal waters, and in fish and shelfish living there. In some areas, mercury levels in species such as tuna were high enough to make these fish unsuitable as human food, (Perhaps the most notorious incident of all is the outbreak of "Minamata Disease" in Japan.

Methylmercury discharged from a chemical factory into the sea, or produced in the sea from inorgnic mercury discharges, contaminated the fish eaten by local people at the town of Minamata of Kyushu island, Japan. As a result of this and a similar incident at Nilgata on the east coast of Honshu, nearly two thousand people have so far suffered neurological disorders; about four hundred have died). Some of the pollution came via river; the amount of iron, manganese, copper, zinc, lead, tin and antimony that reached the sea by this route was far greater than would be supplied by natural geological processes. Other contaminants came through atmospheric deposition: the importance of this pathway for metals and synthetic chemicals was increasingly recognized during the decade (1972-82). Offshore oil and gas exploration and dredging for sand and gravel in coastal areas increased during the period 1972-1982. Coastal zone development affected extensive estuarine areas, as well as mangrove swamps and coral reefs. Oil pollution killed sea-birds fouled beaches and affected tourism.

Although tanker accidents were the source of less than 5 percent of all the oil entering the sea, accidents released large volumes in small areas, and were therefore especially damaging. A recent report issued by a group of experts on marine pollution and published by the United Nations Environmental Programme, Warns that increasing contamination of the coastal seas, which has so far only caused localized disturbances to the marine environment may become much more serious. The combined effects of the dumping of sewage sludge industrial wastes, litter and oil spills into the Mediterranean, the North Sea, the Baltic and the Gulf of Mexico, may well cause ecological disasters which could gradually lead to a global crisis since the oceans are a unified life support system.

The report points out that the sea's capacity to absorb wastes has its limits, so that as the world economy continues to grow, even more sewage, industrial wastes and other effluents are likely to find their way into the sea. Man-made radioactive materials have been discovered in the open ocean and scientists have found traces of DDT in th Arctic sea and the Antarctic penguin, thousands of miles away from the areas where the pesticide is used. And because many of the disturbances which are introduced occur at vulnerable points such as the surface of the sea or on the sea floor, where oceanic life-supporting processes are generated, the dangers and magnified.

Although the modification of this microlayer has occurred only on the vicinity of oil spills and around the marine dump site of municipal wastes and no evidence has yet been found that detergents in the open ocean have significantly influenced the properties of the sea surface, the report points out that environmental catastrophies in certain coastal areas and seas might already have occurred.

Only improved management of the coastal areas and seas which involves controlling more closely the dumping of pollutants and careful monitoring of contaminant levels, will prevent a further deterioration of coastal environments. Much work however is still to be done before the several ecological sub-systems which constitute the ocean are fully understood. By the end of 1982 action plans were in operation or planned for 10 regions involving some 100 coastal States. Coastal and marine national parks and reserves were extended.

(e) Desertification

Desertification is now perceived as an intensifing world wide threat which, to varying degress, affects over 100 Member States of the United Nations. More than 20 percent of the world's population (850 million people) live in the 35 percent of the Earth's land surface (45 million km) which consists of the arid, semi-arid and sub-humid zones which is at risk of desertification.

Three quarters of this area is already at least moderately desertified. Each year an additional 21 million hectares of agricultural land deteriorates through desertification to a point at which it is no longer economically productive. The last production has been valued at 26 billion dollars annually. Although desertfication is exacerbate by severe drought, its principal cause is human over exploitation of dry lands through over-cultivation, overgrazing, poor irrigation practices and deforestation. When drought strikes the overtaxed livelihood systems, desertification results. The most devastating environmental and economic effects of desertification are in developing countries in the drylands.

National balance of payments problems and local rural poverty lead to over-exploitation of lands and forests for food, fuel, cash crops and meat for export. As these and other pressures increase to unsustainable levels, agricultural productivity and soil fertility decline, creating in turn new pressures which lead to further degradation. Recovery of land for agricultural purposes become costly and often economically unfeasible, thereby reducing the economic potential of the affected countries. In many of these countries agricultural land is also being reduced by salinization, urban sprawl and sand-dune encroachment.

Increased desertification in one part of the world can indirectly induce desertification else where, as loss of agricultural land and increased food demand raise Prices and intensify pressures on marginal drylands in other food exporting and importing countries alike. There are also other impacts on international trade, to the extent that developing countries affected by rapid desertification will have a reduced capacity to import goods and services from developed countries and from other developing countries.

Other globally shared impacts include loss of valuable genetic resources, disruption of hydrological cycles and increase of atmospheric dust. A great deal is already known about the causes and effects of desertification and about the measures needed to halt or combat it. Moreover, there is favourable cost benefit ration to justify, preventive action. The 1980 estimate for implementing the 1977 plan of Action to combat

Desertification was 2.4 billion dollar annually over 20 years, which is only a fraction of the estimated value of agricultural production lost each year through desertification. It is, therefore, in the interests of all countries to establish some concreate cooperative actions between groups of developed and developing countries to halt further desertification, a goal which can be achieved, given a sufficient allocation of resources, on the basis of currently available knowledge.

(f) Loss of Tropical Forests

Although tropical forests are found in early half of all United Nations Member States, most are in twenty countries and nearly half their total area in three countries alone (Brazil, Indonesia and Zaire). The tropical forests are the world's richest biological zones, and are estimated to contain as much as 40 per cent of all terrestrial species on the planet. In addition, tropical forests provide a wide range of useful products of economic significance for both developing and developed countries.

Undisturbed tropical forests are also home to millions of the world's tribal peoples. Many experts contend that tropical forests are being exploited at a rate and in a way that is ecologically destructive and economically unsustainable. There are varying estimates of how much of the tropical forests will be lost by the year 2000 through complete conversion to other uses and by severe degradation. The results of the tropical forests resources assessment project conducted by FAO and UNEP indicated an estimated minimum loss of closed tropical forests of 12.5 per cent by the year 2000 (an average of 7.5 million hectares annually between 1980-2000) through complete conversion to other uses alone. Other studies which not only consider complete conversion but also include severe degradation caused, for example, by logging operations, by over-grazing, by over-exploitation for fuelwood, or by natural causes such as fires, insects and diseases, have estimated losses of tropical forests as upto 20 million hectares or more in a year.

The FAO/UNEP study estimates that tree planting on industrial and non-industrial plantations accounts for only 10 per cent of the area of tropical forests cleared annually. The major causes of deforestation differ by region. According to the FAO/UNEP report in Latin America the two major causes are shifting cultivation and the conversion to pasture land for cattle ranching. In Asia and Africa, shifting cultivation by subsistence farmers and removals for fuelwood are major causes. The extraction of timber from tropical forests for industrial purposes is not considered by the FAO/UNEP reports as a major factor in deforestation.

On a world-wide basis, the UNEP report estimates that fuelwood and charcoal removals from tropical forests are as much as eight times greater than for industrial wood. Serious degradation of loss of watershed forests in particular has a wide range of major ecological and economic effects through increased erosion, floods, landslides, and silting of hydro-electric facilities, irrigation systems reservoirs and harbours. The lives and livelihood of half the world's population depend directly on the wise management of watershed forests and. eco-systems.

Deforestation also influences local, regional and possibly even global climatic conditions by altering hydrological regions, reducing vegetation cover and causing more heat and light to be reflected back to the atmosphere (the albedo effect), and through accelerated burning of the forest biomass, which contributes to carbon dioxide build-up in the atmosphere and the global warming trend. The economic, social and political implications of possible regional and global climate changes have only begun to be assessed.

(g) Soil Loss

Almost every environmental concern connects clearly with the importance of soil. Soil acts as a reservoir for water and other inorganic substances cycling through the eco-system. It provides habitat for many of the different populations or decomposer organisms that control the rate and capacity of nutrient recycling.

The importance of soil to agriculture can hardly be over emphasised. Agriculture provides about 44 percent of the poorest countries gross domestic product and 83 percent of their employment. Growing more food, building irrigation projects, conserving nature and precious wildlife and other needs must depend on proper soil management. The wide-ranging importance of soil confirms that soil management must consequently be many-sided. The broad range of natural and man-made stress that causes the two type of soil damage: physical displacement and alterations in soild structure and composition.

Uncontrolled shifting cultivation, salinization and water-logging flooding and plant cover destruction are same. An appropriate use of chemicals, weapons testing and deployment in warfare are some other casuses of soil nutrients and other types of soil degradation. Soil damage is also caused by patterns of dry land mountain side cultivation that carried out without arrangements for bunding and terracing to protect the topsoil.

Continuous cropping to meet growing food needs leave title time for soil conservation techniques. Deterioration of the productive capacity of the soil is most acute in areas affected by erosion, drought, mineral stress and other related factors. Soil resources that are vulnerable to these factors constitute 82 percent lands in South Asia. While erosion is an integral part of the natural system, occurring even when land is covered by grass, or trees, its extent and intensity can be minimized provided land denudation is correspondingly reduced.

The fact that it takes nature 100 years or more to build one inch of top soil should have engineered the vigilant observance of soil conservation measures. Instead, problems of soil erosion have continued ticking. Countries in East and South Asia are particularly burdened with soil erosion occuring in watersheds.

Anticipated acceleration of agricultural production and other activities between now and 2000 is casting its shadows: existing cultivable land will be further subjected to intensified use as cultivation of unsuitable areas persists; logging, firewood gathering and grazing will accelerate the elimination of vegetative cover, while the construction of roads, mine and water resources projects will be further pursued in the course of development. Such prospects place the least developed countries in an even more precarious situation.

Slash and burn farming is a common enough culprit. Wielding destruction in gradual succession, it initially causes the loss of fertility by the thin topsoil, and later eroding its stability. In India, slash and burn farming is particularly rampant in the north-eastern and central regions affecting between six and eight million hectares of land. Salinization affects 80 million hectares (8 percent) of total land in South Asia.

Something should also be said about agricultural techniques which are designed to raise land productivity but instead lead to soil loss. The use of cheap nitrogen, for example, replaces nitrogens, fixing legumes, resulting in loss to topsoil.

The commonality of problems, in addition to the implications transboundary soil management points up to the fact that soil problems cannot always be resolved within the context or national concerns. Clearly, what are required are mutual policies and strategies concerning land management, and the World Soils Policy provides an appropriate starting point. Working in collaboration with the relevant UN bodies, UNEP has proposed the World soils Policy programme which endeavours to enhance international awareness, assist countries in the formulation of national soil policies, help develop technical and scientific knowledge and collect, compile and disseminate data on the use and management of the worlds soil resources.

(h) Depletion of genetic resources

25,000 species and more than 1,000 vertebrate species and sub-species are now estimated to be threatened with extinction, yet these two categories of organisms represent only a small proportion of the terrestrial species on the planet. According to some estimates as much as from half a million to a million species would be extinguished over the next two decades. Extinction of species on this compressed scale would be without precedent in human history. As much as two-thirds of all terrestrial species and the majority of endoangered species are located in developing countries.

A significant loss of species in developing countries would have a world wide effect on future agriculture, industry and science, and on human health and welfare generally. The most serious threat to many species is the disruption and destruction of their habitat by man, especially in the tropical forests which alone harbour as much as 40 per cent of all terrestrial species. Significant species loss could directly affect human health and welfare locally, and even nationally, in many developing countries, where wild plants and animals are a source of fodder, traditional medicines and income especially for the poor in rural areas. In some countries tourism largely on wildlife is a major source of foreign exchange.

In many developed countries over one-thrid of the gain in the productivity of many major agricultural crops over the last few decades has been attributed to breeding programmes using genetic material from related wild plant varieties, many of which are found in developing countries. Moreover, many of the major crops in developed countries now have a limited genetic base. For example, more than half the Canadian prairie wheatlands are devoted to a single variety, Neepawa, which was developed with germplasm from Kenya; 72 per cent of United states potato production depends on only four varieties; and the entire United States soyabean industry is derived from a mere six plants from Asia.

A significant loss of relavant wild species could limit the possibilities for maintaining or increasing the yields of major crops, such a wheat, maize, rice potatoes and for combating new strains of diseases or pests which threaten them. Genetic stocks from one part of the world frequently make an enormous contribution to another. Current research shows that enormous opportunities lie ahead, such as the utilization of Asian blue-green algae of fix nitrogen for crop fertilization, which may constitute a gain of hundreds of millions of dolars a year to the economics of developed countries alone.

Major medicines and other pharmaceutical products in developed countries are also often dependent on plant and animal species. For the United States of America it has been estimated that more than 40 per cent of the prescriptions written each year contain a drug of natural origin-from higher plants (25 per cent), microbes (13 per cent) or animals (3 per cent) as the sole active ingredient or as one of the main ones. It has been estimated that among the 10 percent of species that might be extinguished by the year 2000 there could be many plants with potential for use in new plantbased drugs and pharmaceuticals with a combined world wide commercial sales value of as much as 200 million pounds annually. Biomedical research in developed countries is also vitally dependent on wild species, many of which are threatend with extinction.

The innovative capacity and future of the rapidly expanding field of bio-technology, practised largely in developed countries, relies considerably on genetic patterns and resources from many wild plants and animals as only about 10 per cent of the world's plant species and 1 per cent of the animal, species have so far been subjected to even a preliminary screening for potential medical or commercial applications, a significant loss of species could also represent a permanent loss of sources for more effective chemicals, medicines and other useful products. In spite of this interdependent relationship, species loss continues to accelerate.

Developing countries often lack the capacity and economic incentive for extensive species protection. They have derived little or no economic benefit from the development and use of genetic materials in developed countries, where the scientific and technical expertise to exploit species and their genetic resources predominantly resides. Nevertheless, developing countries are and will remain the principal source of genetic materials from wild plants and animals. This clearly shows that there is a common incentive for developed and developing nations to prevent a potentially catastrophic loss of species.

The mutual advantage of comprehensive joint effort could result in new medicines and other useful products that would contribute to improved levels of human health and welfare in all countries.

New Approaches to Environment and Development

Human progress has always depended on our technical ingenuity and a capacity for cooperative action. These qualities have often been used constructively to achieve development and environmental progress: in air and water pollution control, for example, and in increasing the efficiency of material and energy use.

Many countries have increased food production and reduced population growth rates. Some technological advances, particularly in medicine, have been widely shared. But this is not enough. Failures to manage the environment and to sustain development threaten to overwhelm all countries. Environment and development are not separate challenges: they are inexorably linked. Development cannot subsist upon a deteriorating environmental resource base: the environment cannot be protected when growth leaves out of account the costs of environmental destruction. These problems cannot be treated separately by fragmented institutions and policies. They are linked in a complex system of cause and effect.

First, environmental stresses are linked one to another. For example, deforestation by increasing run off, accelerates soil erosion and siltation of rivers and lakes. Air pollution and acidification play their part in killing forests and lakes. Such links mean that several different problems must be tackled simultaneously. And success in one area, such as forest protection, can improve chances of success in another ares, such as soil conservation. Second, environmental stresses and patterns of economic development are linked one to another. Thus agricultural policies may lie at the root of land, water, and forest degradation.

Energy policies are associated with the global greenhouse effect, with acidification, and with deforestation for fuelwood in many developing nations. At these stresses all threaten economic development. Thus economics and ecology must be completely integrated in decision-making and law making processes not just to protect the environment, but also to protect and promote development.

Economy is not just about the production of wealth, and ecology is not just about the protection of nature: they are both equally relevant for improving the lot of humankind. Third, environmental and economic problems are linked to many social and political factors. For example, the rapid population growth that has so profound an impact on the environment and on development in may regions is driven partly by such factors as the status of women in society and other cultural values. Also, environmental stress and uneven development can increase social tensions. It could be argued that the distribution of power and influence within society lies at the heart of most environment and development challenges. Hence new approaches must involve programmes of social development, particularly to improve the position of women in society, to protect vulnerable groups and to promote local participation in decision making.

Finally, the systemic features operate not merely within but also between nations. National boundaries have become so porous that traditional distinctions between matters of local, national, and international significance have become blurred. Ecosystems do not respect national boundaries. Water pollution moves through shared rivers, lakes, and seas. The atmosphere carries air pollution over vast distances. Major accidents—particularly those at nuclear reactors or at plants or warehouses containing toxic materials—can have widespread regional effects.

INDEX